For my friends

Marilyn MONROE

GRAHAM McCANN

Rutgers University Press
New Brunswick, New Jersey

First published in the United States of America in cloth and paperback by
Rutgers University Press, 1988

First published in Great Britain in cloth and paperback by Polity Press, in
association with Basil Blackwell, 1988

Library of Congress Cataloging-in-Publication Data

McCann, Graham, 1961–
  Marilyn Monroe : the body in the library

  Bibliography: p.
  Includes index.
  1. Monroe, Marilyn, 1926–1962. 2. Moving-
picture actors and actresses — United States —
Biography. I. Title.
PN2287.M69M33 1988 791.43'028'0924 [B] 87-26522
ISBN 0-8135-1302-2
ISBN 0-8135-1303-0 (pbk.)

Typeset in 10 ½ on 12pt Sabon
by Opus, Oxford
Printed in Great Britain

Contents

Acknowledgements

Many people are associated with the writing of this book, whether through their encouragement, intellectual advice or friendship, good humour and support. John Thompson made himself available to comment on my ideas at every stage of the work. I learned much from conversations with John Dunn, Ernest Gellner, Irving Louis Horowitz, Alice Jardine, Ian Jarvie and David Reason. Anthony Giddens showed himself to be not only a good critic but also an eminently kind friend. Amongst those scholars whom I know only through their work, Stanley Cavell and Richard Dyer have helped me with their excellent and thoughtful studies.

Several people kindly responded to my queries with advice on specific points: Saul Bellow, Alistair Cooke, Jack Lemmon, Norman Mailer, Arthur Miller, David Quinlan. W. J. Weatherby has been a very helpful adviser; I greatly appreciate his sensitive concern for the memory of Monroe. The staff of the British Film Institute were always helpful and highly efficient (the projectionists in particular were welcome companions during my increasingly lonely viewing of Monroe movies!). I must also thank the staff of the Margaret Herrick Library of the Academy of Motion Picture Arts and Sciences, and the American Film Institute in Hollywood and in Washington, D. C. The organizer of the Marilyn Monroe Fan Club (UK), Stewart Hutcheon, generously gave me access to numerous papers and letters. Mrs May Ross permitted me to read her interesting correspondence with people connected with Monroe.

My personal debts are even more extensive and pleasing to record. I cannot adequately thank my parents for their constant encouragement and faith in me. I have had the good fortune to test my ideas in relaxed conversation with the following friends: Molly Andrews, Stephen Bann, Norman Bryson, Judith Ennew, Anna Grimshaw, Stephen Gundle, Annie Lycett, David McLellan, Marissa Quie, Shirin Rai, Amanda Reiss (as ever), Ingrid Scheibler, Alex Schuessler, Louise Simpson, Marion Smith, Richard Sparks, Wesley Stace and Alison Young. Teresa Brennan discussed with me some of the important points in the text. Christopher and Linda Taylor have been extraordinarily loyal supporters and sympathetic critics – they reaffirmed my faith in the possibility of combining intellect with genuine compassion. ('The heart', as Woody Allen says, 'is a very, very resilient little muscle.')

Silvana Dean prepared the manuscript and tolerated my frequent revisions; to do so with such constructive grace made my own work far less taxing. I make one special dedication: it is for my niece, Penelope (born as this book was being written) – I wish her a wonderful future. The text was written especially for my friends, forever in my thoughts, always on my mind. The mistakes are all mine, what remains has been touched by their presence.

Graham McCann
Cambridge

Introduction

Please don't make me a joke.

Marilyn Monroe

Please do not understand me too quickly.

André Gide

In 1962, shortly before she died, Marilyn Monroe was interviewed by Margaret Parton for a special feature in the *Ladies Home Journal*. The star of *Some Like It Hot* and *The Misfits* had recently been divorced from Arthur Miller and sacked by her studio. Rumour had it that Monroe was dependent on drugs, drifting through life, too famous to fade away but also too famous to start on a new direction. Parton arrived at Monroe's residence with these rumours in mind. An assistant came to greet the reporter, asking her not to question Monroe about the past – 'She's looking forward, not back.' The interview was a revelation, Parton finding herself conversing with a lively, intelligent person, evidently full of hope and determination and with an expressed desire to 'put the record straight'. Reflecting on the meeting, Parton composed her thoughts: 'beneath the sickness, the weakness and the innocence, you find a strong bone structure, and a heart beating. You *recognize* sickness, and you *find* strength' (in Guiles 1971: 304-5). Here was no 'ageing sex symbol', this was not a person caught up in some 'inevitable' decline: Parton had spent several hours with a talented woman who was busily planning her future. The article was written with this impression at its centre. It was never published. Parton's editors, Beatrice and Bruce Gould, informed her that she had been far too sympathetic – she had clearly been 'mesmerized' by Monroe. Bruce Gould took Parton to one side and explained: 'If you were a man, I'd wonder what went on that afternoon in Marilyn's apartment.'

'People didn't take me seriously,' complained Monroe, 'or they only

took my body seriously.' Now, more than a quarter-century after her
death, many people still do not take Marilyn Monroe seriously. 'Please
don't make me a joke...' The treatment of Monroe, by critics,
biographers and admirers, is still an *issue,* it still stirs controversy.
Compare the following two evaluations of Monroe – the first by movie
director Joshua Logan, the second by journalist Val Hennessy:

> It is a disease of our profession that we believe a woman with physical
> appeal has no talent. Marilyn is as near a genius as any actress I ever
> knew. She is an actress beyond artistry. She is the most completely
> realised and authentic film actress since Garbo. She has that same
> unfathomable mysteriousness. She is pure cinema.
>
> (In Taylor 1984)

> Marilyn was a hopeless actress, a self-indulgent slut, a weak-willed
> wimp who dressed up like a female impersonator, and a star whose
> teasing, titivating, feather-brained act, on and off screen did nothing
> for the dignity of women.
>
> (*Daily Mail* 5 March 1987)

Monroe once observed that she had never 'met a writer I'd like as my
judge'. She has since been served up for judgement in over fifty
biographies, picture books and documentaries, countless articles, endless
theories. When no more images seem left and no new anecdotes appear
available, investigative journalism digs deeper and emerges with further
intrigue and fresh controversy. The image 'Marilyn' leaves behind the
individual Monroe, like shedding a skin. This 'Marilyn' is copied,
constructed, or created – a *man*ufactured Marilyn; each image means,
implies, reveals, connects; each instance is eventually either consumed or
conserved. The person may be dead but the myth has never been more
alive: who can get through one week without seeing a poster of Monroe in
her white dress, or that laughing portrait on a greetings card, or those
books about her, or a remark on television or in a conversation: 'That's
like Marilyn,' 'Monroe was like that.' Her portrait has been pressed into
an endless routine by Andy Warhol's silk-screen prints. Her figure is
reflected in the 'Marilyn' effigy in California's Movieland Wax Museum
at Buena Park. She has been reproduced, reconstituted, *rewritten.* The
effect this process has on her image, on our memory of Marilyn Monroe,
is the subject of this book.

Monroe's life and career has often been viewed through the flickering
light of Hollywood and the movies. Certainly, Hollywood was one of the
most symbolic, emotionally valent landscapes in the USA, being a

montage of ideas – a town, an industry, a frame of mind, a self-actualizing myth. Hollywood had Stars, always capitalized, the 'S' shooting up like a skyscraper. 'MM' was the construction Monroe was obliged to inhabit, but as she said at the end of her life: 'that's not where I live.' Since her death, there has been an unparalleled number of hardened movie figures who have recorded how profoundly they miss her. Many studies of Marilyn Monroe purport to uncover 'the *real* Marilyn', the person beneath the layers of rumour and hearsay, the truth behind the fictions – the body in the library. Indeed, since the recently renewed interest in the mysterious circumstances surrounding her premature death, Monroe's writers have come to resemble detectives involved in a celebrity 'whodunit'. Any man who writes today on Monroe (and, typically, it is men who do – who are allowed to) faces two fundamental problems of distance: the distance between men and women, and the distance between the researcher and the deceased subject. It is these problems in particular that inspired the present study.

When the death of Monroe was announced by the *Los Angeles Times* (6 August 1962), a banner across the front page read: 'Star's Life in Photos, Stories'. How can her life still *be* in 'photos, stories'? Why do we read her biographies and gaze at her image and, at times, feel that she is 'here' and that we 'know' her? Monroe remarked that, 'Those who know me better, know better.' How can we now try to know her better? The difficulty is compounded by the elusive nature of the personality of a performer: Monroe was a professional role-player, she assumed a multiplicity of selves, of 'Marilyns', making it hard even for her to feel settled within any single identity. As a celebrity she drew out the actor in people; she said that fame 'creates a self-consciousness in other people toward you. . . They become actors and characters rather than behave as they normally would.' Thus, the attempt at 'finding *the* Marilyn Monroe' is a very frustrating project. My early chapters on biography, pictures, movies and fandom will try to show this. What becomes clear, I believe, is the need to comprehend the complexity of Monroe – and this entails comprehending the complexity of *ourselves*. There was never a clear and compact self called 'Norma Jeane' that was later dropped into a volatile environment and dissolved like a soluble pill into a thousand 'Marilyns'. There is never a static, consistent self called the Author, who sits in judgement over Monroe. Chapters 6, 7 and 8 highlight this fact, this awareness of difference: by dealing with Arthur Miller and Norman Mailer as they try to 'capture' their ideal 'Marilyn'; by reflecting upon the desire in many of us to retain a sense of her presence after her death; and finally by considering my own relationship with someone I will never know.

Above all, I hope to explain the myth of Marilyn Monroe in such a way as to reveal the difficulties in interpreting the public image, whilst retaining the desire to understand the person. The 'dumb blonde' image is *there,* on that poster and that print; the intelligent individual is also evident, for those prepared to look. Does the latter cancel out the former, or vice-versá, or does the biographer refrain from any judgement? This book is following in a daunting tradition: so much has been written, so much has been repeated. The remark by Gide (1891) seems peculiarly pertinent here: 'Everything has been said before, but since nobody listens we have to keep going back and beginning all over again.' Not quite all over again: I am writing within the tradition but also *against* it, against the existing representations of Monroe by other men. In this endeavour I have embraced the advice by Showalter (1983/4: 143): 'The way into feminist criticism, for the male theorist, must involve a confrontation with what might be implied by reading as a man, and with a questioning or a surrender of paternal privileges.' Thus, as I interpret Marilyn Monroe, I must also interpret myself. She said that it saddened her how writers often 'observed' people without 'feeling their feelings'. I share her sentiment. In what follows I hope I have made this evident.

I

The Myth of Marilyn Monroe

*It is so difficult to find the beginning, or, better: it is difficult to begin
at the beginning. And not try to go further back.*

Ludwig Wittgenstein

*I want to grow old without face-lifts. They take the life out of a face,
the character. I want to have the courage to be loyal to the face I've
made. Sometimes I think it would be easier to avoid old age, to die
young, but then you'd never complete your life, would you? You'd
never wholly know yourself.*

Marilyn Monroe

*It stirs up envy, fame does. People you run into feel that, well, who is
she – who does she think she is, Marilyn Monroe? They feel fame
gives them some kind of privilege to walk up to you and say anything
to you, you know, of any kind of nature – and it won't hurt your
feelings – like it's happening to your clothing.*

Marilyn Monroe

Virginia Woolf wrote that 'A biography is considered complete if it merely
accounts for six or seven selves, whereas a person may well have as many as
one thousand.' Marilyn Monroe always remembered this comment; her
biographers rarely do. During her life she observed: 'I carry Marilyn
Monroe around me like an albatross' (in Weatherby 1976: 45). How do we
understand Marilyn Monroe? Should we line up the images and interpretations like a school assembly, or sift through and separate wheat from
chaff? Do we know more as we uncover lost documents and long-guarded
photographs? With Monroe, the prints have been over-exposed and the

prose seems to pale in comparison to her own, fleeting exclamations. Monroe's influence on popular culture is still enormous: her image, her performance, her personality, her style, her secrets, her story, are still affecting us. We still want to know Monroe, yet we are still unsure what this means.

Saul Bellow described popular culture as a 'Moronic Inferno', a world characterized by changing characters, by greater power and less control, by more systematic intelligence and less searching intellect. It is in such a milieu that Marilyn Monroe came to prominence. The study of her image and its impact upon the famous few and the unfamous majority can provide us with a fascinating insight into the tensions between public and private lives, between performance and its presentation, and the process whereby the play of celebrity symbols largely overtakes the play of ideas in social experience.

> [A] sex symbol becomes a thing, I just hate being a thing. But if I'm going to be a symbol of something I'd rather have it sex than some other things we've got symbols of.
>
> Marilyn Monroe

Norma Jeane Baker escaped from an early life of lovelessness and foster homes to become a celebrity, 'Marilyn Monroe', a cultural phenomenon. On 5 August 1962, when Marilyn Monroe's housekeeper found her naked in bed, killed by an overdose of barbiturates, it was the end of Monroe, the beginning of a spectacular script for *Marilyn's* entry into myth. The person has died, but the myth is alive and flourishing. Behind the hyperbole and the hypotheses there was an orphan who became the most famous woman in the world, a child deprived of much maternal encouragement who grew up in a paternalistic milieu, who was labelled with love whilst forced into loneliness, who died publicly but in pain at the age of thirty-six. She was marketed as the modern mistress, yet she yearned for monogamy and motherhood. The profile was cast as crude whilst the passion was for culture. The genius of the performer masked emotions marked with pain and insecurity. The significance of her sexuality and her success has stimulated a succession of interpretations in a variety of media. Monroe's photographer friend, André de Dienes, recalls the first few hours in which Norma Jeane Baker, encouraged by her movie studio, began rewriting her self and her signature as 'Marilyn Monroe'. According to de Dienes (1986: 78), Monroe 'had picked up a pencil and was trying out her signature with two large, curly, romantic Ms on a notepad. She was getting acquainted with her new identity, saying "Marilyn Monroe" as if tasting a piece of candy.' 'MM' – man-made; 'MM' – modern myth; 'MM' – *memento mori*. We are now on first-name terms, good terms, our own terms, we now feel familiar with the person and the memory she invokes.

The circumstances of her death – the drug overdose, the implications of powerful figures, the frustrated attempts to find the desired voice on the other end of the telephone – belong, with tragic appropriateness, to the B-movie script. Monroe was as much a creator of the movie capital as it was of her: as she said, 'Los Angeles was my home. . . so when they said, "Go home!" I said, "I am home".' Hollywood's own sweetly cynical screen writers might have killed her in the same way on the screen, just as her life had become an intricate tangle of public and private appearances, an exploitation of personality and at the same time a submerging of identity into whichever image of 'Marilyn' the public wanted. Monroe had the quality of vulnerability: she looked as though she could easily be hurt. She was a star as well as an actor (the term I shall use to refer to both male and female performers): on the screen, she succeeded in stripping away the top layer of personality, encouraging us to believe that we were seeing through to some essential quality of love or loneliness, some private pleasure or pain. In contrast to the indestructability of a Joan Crawford or a Katherine Hepburn, Monroe seemed to belong definitively to the uneasy, nervous Hollywood of the 1950s.

Monroe's death was followed by a wave of frustrated pity, a sort of shared guilt summed up in Laurence Olivier's comment that she was 'exploited beyond anyone's means'. Part of Monroe's appeal was to the audience's protectiveness; part of her posthumous appeal is to the biographers' protectiveness. If Monroe's death symbolizes anything, it is disenchantment: the slow, soft, sad fading of a dream, and one dreamt not only by the star ('It's nice to be included in people's fantasies, but you also like to be accepted for your own sake'), but also by the industry and the public which courted her. Her posthumous status sadly reflects the still strong influence of sexist ways of thinking: 'take her seriously' is still an often-ridiculed request. Charles Chaplin was a comedian who yearned for culture and the concomitant civilities of the artistic elite although his quest for self-improvement was arguably no more worthy (or more successful) than Monroe's, yet his biographers consider the quest with considerably more respect.

It is Monroe's symbolic value that has become uppermost in the posthumous phase of her image's career. Hembus (1973: 12) catalogues something of the range of uses her image has been put to: 'The worth or worthlessness of commercial Hollywood, intellectual New York, the Press, psychiatry, religion, cultured Europe, barbaric America, liberation, masculine chauvinism, the Establishment, the anti-Establishment, manipulation and, finally, the quality of the American dream and the Kennedy family's mode of behaviour.' We can thus see three recurrent emphases: Hollywood (symbolizing either capitalist America or the mass media), women, and sex. Monroe as a movie star represents those who believed Hollywood's dreams, and the destructive force of the 'dream-factory'. Monroe, as a female star, seems to dramatize the conflict within women between the opposing pressures to be an object for men and a subject for themselves. Monroe, as a sex symbol, has attracted perspectives of male heterosexual response to the image: very rarely do writers consider what it means for a woman to embody this symbol, nor for other women to have to come to terms with this embodiment. Monroe's posthumous charisma can be considered in relation to the flux of ideas about women and female sexuality occasioned by the insights of feminism and the responses, accommodations and resistances of the mass media to them.

Many people may be surprised that there are any real secrets left to be told about Marilyn Monroe. Yet the many layers of literature on the woman have only made her more distant, the interpretations being rich in intrigue but poor in insight as Monroe dissolves in a mist of fresh metaphors and stale anecdotes (the latter mostly borrowed and mostly blue). Along with Bogart, Dean, Garland, Presley and Lennon, Monroe's post-mortem popularity is as great, if not greater, than it was when she was alive. More so than any other celebrity, Monroe became a myth which was projected and processed through Warhol prints and Washington inquests, a memory marred by the media.

A succession of authors and artists have responded to Monroe and her story. At one extreme it triggered the tricksy pretension of the Nicolas Roeg film *Insignificance* (1985); at the other, it inspired Norman Mailer's 'novel biography' of a 'possible' Marilyn, which proved a triumph of intuitive speculation. A recent biography by Anthony Summers, *Goddess: The Secret Lives of Marilyn Monroe* (1985), was inspired by the 1982 reopening of the case concerning Monroe's death, and it includes material collected from the testimonies of six hundred people. Contrary to appearances, this literature on Monroe has not reached saturation point. *Goddess* represents the opening up of the subject, being more interesting for the questions it raises than for the conclusions it contains. In terms of style, Summers signally fails to avoid the formidable historiographical and ethical problems involved in the interpretation of such a complex and controversial career. In terms of content, he has sparked off new speculation concerning the role of the Kennedys and the Mafia in Monroe's final years, and a public inquiry has once again become a possibility. The demand for further discussion and debate is more intense than ever.

The predominant conceit is for biographical accountancy, wherein each fact is a figure in a column which can be added up, 'settled', recorded. What these biographers have failed to do is to pierce through the layers of interpretation and uncover the 'authentic' person. What they have succeeded in doing is in contributing to the Monroe myth, actively adding to the distance that separates us from the real person, drawing us deeper into the cultural mythology. When we now think of Marilyn Monroe, we summon up a montage of memories stretching across space and time: the blonde woman with her white skirt blown high around her waist; the gaudy images of a Warhol print; the Monroe mimics in movies and advertisements; the smiling face on the shiny greetings card; the suggestive pose on the poster; the 'Marilyn' of the memoir, the memorial, and the museum. Although my own approach incorporates a biographical aspect, my prime concern is more distinctive.

The mediating process that links 'bios' with the 'graphic', which translates and transcribes a life on to paper and print, is an eminently intricate and interesting activity. My project is thus inspired by the problems of interpreting a career which involves the media of film, photography, music and literature; a life which has become so obscured by exploitative journalism, inconsistent use of primary sources, and unreliable testimonies and lamentations, that any biographer is more concerned with previous biographies than with the 'subject' as such; and a myth which is largely man-made, mass marketed, and mass consumed. In the aftermath of Monroe's death, the initial opposition between a career and its various stages assumes another kind of illusory completedness, a unified whole prefigured as a woman covered with an endless white garment, a weaving of body and garment that fulfils the infinite text produced in and through reading. The nature of this cultural myth will be explored, both for its constraining and its enabling functions.

Marilyn Monroe epitomizes the ambiguities in modernity. Reactions to her ubiquitous image, at once so touching and so tacky, reveal the tensions in positions which long for transcendence whilst relishing tumult in a society lurching from fashion to fashion, force to force. According to Walter Benjamin, in modernity one can never finish what one has started; the figures of the labourer and the gambler represent the disintegration of coherent experience. The labourer and the gambler are caught up in all-consuming activities, unable to 'complete' the game. Benjamin held the conviction that every cultural product experiences an autonomous 'post-history' by virtue of which it transcends its determinate point of origin: 'The work is the death-mask of the conception.' My concern with the post-history of Monroe is in the spirit of a 'redemptive critique', an analysis aiming not at a negation of the contents of a tradition but rather an appreciation of that tradition and its relevance to the present day. My interest is with the tensions within the conventional image of a star, a celebrity from a medium that has made celebration into an art of sorts. To analyse celebrity is to bring out the disjunction between what its image says and what its figural workings constrain it to mean.

Not only is the modern celebrity visible, she or he has the power to 'be' everywhere at once: a technological solution to the restlessness that marks the Western character. 'Power passes so quickly from hand to hand,' de Tocqueville wrote of America, 'that none need despair of catching it in turn.' Andy Warhol pronounced, 'In the future everybody will be famous for fifteen minutes.' Senator McCarthy fueled fifties paranoia with a flurry of unfounded accusations concerning communism; sixties hippy-radical Jerry Rubin accepted the 'Academy Award of Protest'; when the American hostages were finally returned from Iran early in 1981, a State Department spokesperson said that they 'would be free either to cooperate with news organizations and become celebrities or to withdraw quietly into private life'. John W. Hinkley cited a movie, *Taxi Driver,* as the inspiration of his 'historic deed' in shooting the President. Particularly notable is the case of French bankrobber Roger Knobelspiess, who received a pardon from the Mitterrand government after pressure from the intellectual elite: two years later, Knobelspiess found himself surrounded by armed officers after he had made another robbery. 'Don't shoot,' he cried out, 'take pictures!.' In 1987, an official could summon a press conference and commit suicide on camera.

The history of celebrity and the history of communications technology over the past century are very closely linked (see Quart and Auster 1985). The proliferation of information (made more prodigious and proficient by each new technical development) exacerbated a need for simplifying symbols that crystallize and personify an issue or idea. Ever more briefly as the flow of competing stimuli continues to quicken, these figures help to 'resolve' ambivalence and ambiguity both in the public and private spheres. Concomitantly, each further development in communications has increased our illusion of collusion with the celebrated, bringing proximity with no intimacy.

Stars, Raymond Durgnat (1967 137–8) has said, 'are a reflection in which the public studies and adjusts its own image of itself. . . The social history of a nation can be written in terms of its film stars.' Schickel (1974: 27) has summarized the development of movie stardom as an institution; the process shows us

> how the producers had resisted giving billing to the actors who played in their title films; how the actors themselves, regarding appearance in

a medium that robbed them of what they regarded as their prime artistic resource, their voice, had been glad to hide their shame in anonymity; how the public had begun singling them out of the crowds on the screen, demanding to know more about them, and, more important, demanding to know, in advance, which pictures featured their favourites; how a few independent producers, grasping at any weapon to fight the motion picture trust (composed of the major studios), had acceded to public opinion and had been rewarded by the most deliciously rising sales curves; how the demand for stars was quickly perceived as a factor that could stabilize the industry, since this demand was predictable in a way that the demand for stories or even genres was not; how, as feature-length films established their popularity and the cost of producing these longer films required bank loans, star names came to lead the list of collateral that bankers looked upon with favour when their assistance was sought; how certain actors achieved unprecedented heights of popularity and prosperity almost overnight in the period 1915-1920; and how this phenomenon, this beginning of a new celebrity system, destroyed or crippled almost everyone caught up in it.

'Who killed Marilyn Monroe,' said Sean O'Casey, 'That's a question. That was a tragedy that affected me very much. I hate the idea of Hollywood in which she had to survive' (in Weatherby 1976: 215-16).

The West Coast was an attractive location for the early film-makers, promising them greater freedom from the trust companies and a reliable climate in which to produce weekly motion pictures. The railroad right of way, as one came into Los Angeles in those days, was blanketed with flowers. 'Magic City', it was said, was bathed in fragrant natural glory. In fact, passengers nearing their destination tended to discard the bouquets their eastern well-wishers had given them, and it was the seeds from these flowers that had rooted along the tracks. Marilyn Monroe was brought up in this climate and this culture. She was shown a picture of Clark Gable when she asked the identity of her absent father. Her mother worked within the film industry. She was sent to stay with several film actors and technicians, and when they were unavailable she would lose herself in the local movie theatre. Los Angeles was her locale, Hollywood was her home, movies were her hope.

'Los Angeles,' wrote Reyner Banham 'is the Middle West raised to flashpoint.' Mountains, sea, desert, and an urban area all coexisted within an hour of each other. Climate and topography encouraged immigrants to shake free of their inhibitions and re-create themselves, or lend themselves to re-creation by directors, cameramen, writers and make-up artists. Hollywood had the space in which to flourish, the freedom from the rub of routine reality; it was a place which promised 'overnight success' and wealth in excess, a place to say 'Hooray' for. Monroe (1974: 72) reflected: 'When you're a failure in Hollywood – that's like starving to death outside a banquet hall with the smells of *filet mignon* driving you crazy.'

Hortense Powdermaker, in her 'anthropological investigation' of *Hollywood: The Dream Factory* (1950: 228–9), observed:

> From a business point of view, there are many advantages in the star system. The star has tangible features which can be advertised and marketed – a face, a body, a pair of legs, a voice, a certain kind of personality, real or synthetic – and can be typed as the wicked villain, the honest hero, the fatal siren, the sweet young girl, the neurotic woman. The system provides a formula easy to understand and has made the production of movies seem more like just another business. The use of this formula may serve also to protect executives from talent and having to pay too much attention to such intangibles as the quality of a story or of acting. Here is a standardised product which they can understand, which can be advertised and sold, and which not only they, but also banks and exhibitors, regard as insurance for large profits.

The economic importance of the stars is of aesthetic consequence in such matters as the positioning of spectacle in the presentation of the star, and the construction of narratives which display the star's image. Nonetheless, economics is certainly not the sole determinant of the phenomenon of stardom.

'Only the public can make a star,' said Marilyn Monroe, 'it's the studios who try to make a system out of it.' Such a system was encouraged early in cinema history. D. W. Griffith popularized the use of the close-up as a form of punctuation, a visual variant heightening the dramatic impact of certain moments in certain scenes. There was, however, a more powerful function implicit in the close-up. By moving in closer, Griffith was able to capture a subtler play of emotions on the actor's face, in the eyes. He was trying, he said, to 'photograph thought'. This activity was unprecedented in the history of any art – a breakthrough to (the illusion of) intimacy. Looming over the movie audience came magnified and magnificent images, intricate sections from a larger scene, studied with erotic narrowness and nearness. The close-up enabled the effect of isolating the actor in the sequence, separating the actor from the rest of the ensemble for close individual scrutiny by the audience.

As these images reappeared each week, the individual films began to be perceived (albeit unconsciously) not as discrete creations but as incidents in a more compelling drama – the drama of the star's life and career, the shaping and reshaping of the image that we carry in our minds. Struck by the succession of individual images, the audience starts to see stars. Movie producers began to notice the growing interest in the players who worked for them as reflected in the mail coming to the studios, and they saw the potential value in publicizing them. Actors' names became trademarks for the image: a 'close-up' was affected to produce a sense of propinquity to 'Chaplin', 'Valentino' and 'Fairbanks', eventually reaching an apotheosis with 'Marilyn'. Within four years of this acknowledgement, salaries for leading players rose from $5-15 per week to $250-2,500 per week. Suddenly, salaries were news in themselves, as was the manner in which they were spent, the company the stars kept, the style in which they passed their days. When stars began to make news with their illnesses, with their absence from work, their cultural prominence was assured: 'Marilyn off sick', 'Where is Marilyn?'

The relationship of stars to modernity attracted the attention of social theorists. Herbert Marcuse (1964: 60), sees the 'cultural predecessors' of stars in the 'disruptive-characters [such] as the artist, the prostitute, the adultress, the great criminal and the outcast, the warrior, the rebel-poet, the devil, the fool', but the tradition has been 'essentially transformed'. 'The vamp, the national hero, the beatnik, the neurotic housewife, the gangster, the star, the charismatic tycoon perform a function very different. . . They are no longer images of another way of life but rather freaks or types of the same life, serving as an affirmation rather than negation of the established order.' The example of Marilyn Monroe, as I hope to show in the following chapters, represents the greatest challenge to this viewpoint. Monroe's image was given special attention by her image-makers because she was, for a time, the most marketable star in the world; Monroe herself came to resent the manipulation, and was extraordinarily adept at sometimes subverting her public persona.

The Marcusean approach does not confront the actual content of star images. Examination of these images reveals complexity, contradiction, and difference. 'Audience response' to the projected image is not passive but an active practice, an interpretive process. The close-up, we have seen, reveals the 'unmediated' personality of the individual performer, and this belief in 'capturing' the unique qualities of the performer is probably central to the star phenomenon. As Cavell (1979) observes, this reception is considerably affected by the play of our associative capacities, by our tendency to carry a mnemonic catalogue of previous appearances by particular performers – some correct, some blurred by poor memory, some transformed by fantasy and some a case of mistaken identity.

According to Sam Goldwyn, 'God makes the stars. It's up to the producers to find them' (in Griffith 1970: 25). The modern image does not seek a context; it carries a kind of context within itself, sufficient for its own nourishment, and may work against being placed in history. Monroe's image on a T-shirt or on a postcard seems to animate anonymous raw material. The star's image is apprehended instantly, as a result of the amalgam of old roles, old gossip, old glosses, old publicity photos: such a figure is a walking context, a tradition being transmitted, the face as fashion. In the 1920s, the media had discovered techniques whereby almost anyone could be wrested out of any context and turned into images for the 'silver screen', *Time* and *Life* covers, and the daily tabloids. For no previous era is it possible to make a history out of images; for no subsequent era is it possible to avoid doing so. Countering the festering rumours of illicit sex and drugs in the movie community, Hollywood appointed Will Hayes, the Postmaster General, as head of a new Motion Picture Producer's Association (MPPA), given power to censor on-screen material and to act as moral guardian of the stars' off-screen behaviour. Screen lovers, in their twin beds, had to sleep with bathrobes within a forearm's reach. Kisses could only last five seconds, and lovers must kiss with their mouths closed. By 1930, with power concentrated in the big movie studios, any offender would soon be outside looking in, as Hollywood gave journalists gossip in return for vast discretion when Hollywood required it. The visual and the verbal conversed.

The introduction of 'talking-pictures' radically shortened the psychological distance between stars and their audience. The gods suddenly spoke back to the faithful; the sacred screen image became 'profane'. During the period stretching from the early thirties to the early fifties, a massive effort was made by press agencies to re-educate the movie-going public. A new fiction of 'ordinariness' was fostered, a fiction wherein erstwhile 'extraordinary' people were now presented as entirely like the audience in basic values and harmless vices. Publicity pictures featured the stars hard at work in the studios. Male stars were shown spending leisure time in Hemingway country – at the ranch, hunting, shooting and fishing. (Ronald Reagan appeared to retain this image when, as President of the USA, he wished to be seen 'on vacation'.) Female stars, it seemed, had sufficient spare time to be devoted mothers and endlessly helpful wives. Movie stars were thus portrayed as 'regular guys'. People who possessed more, but nonetheless more of the same. American middle-class normality was, it seemed, worth hanging on to when one's salary started rising.

During World War II, an informal bargain was struck between the US government and the film industry. The government received virtually free, highly professional, assistance in publicizing almost anything it decreed as useful in prosecuting the war. In return, the film industry received favours that ranged from stock footage to delays in the induction of stars who had films to finish. The war gave the entertainment industry the opportunity to extend the long process of 'demystifying' and 'democratizing' stardom: Clark Gable soon to appear in a bomber near you, Errol Flynn sharing a drink with the guys in the canteen.

The traditional task of American historians has been the attempt to authenticate the indigenous mythology by providing it with local material sources: hence Frederick Turner's 'frontier', David Potter's 'material abundance', and Erik Erikson's 'psychological expansiveness'. The film industry's version of this traditional mythology rested on two factors. First, Hollywood's power to produce a steady flow of variations provided the myth with the repetitive elaborations it needed to become convincing. Second, the audience's sense of American exceptionalism encouraged acceptance of a mythology whose fundamental premise was optimistic. To a large extent, American space, economic abundance, and geographic isolation – and the fictions embroidered around these features – had been unavailable to the modern European imagination. We may describe this mythology by observing that, like the invisible style, it concealed the necessity for choice. The consensus contained an underlying premise which dictated the conversion of all political, sociological, and economic dilemmas into personal melodramas – melodramas revolving around the regular, recognizable stars.

The movie-going public did not uncritically assimilate the post-war notion of the Hollywood celebrity as a down-to-earth citizen. The gossip columnists remained powerful opinion prompters: Hedda Hopper, Louella Parsons, Sheilah Graham amongst the Hollywood pack, and Walter Winchell based in New York. These celebrity scribblers supplied a dim, double vision of stardom, a shadowy sense that it was not always as bright and clean and carefree as the picture stories in the fun magazines insisted. A large percentage of the stars still messed up their marriages, made up their biographies, indulged in scandalous affairs, sexual experiments, and alcoholism. So long as the tales from the dark side of the woods did not get out of hand, the underlying 'wickedness' was accepted as an essential element in attracting and holding everyone's attention. *Hollywood Confidential* was an apt title: 'Confession' became an important mode of communication between the public and the private, and the known and the unknown. Epithets such as 'naughty', 'flirty', and 'roguish' took on an endearing quality. Errol Flynn survived a statutory rape charge and was accepted as a 'loveable rogue'. Studios 'protected' their star properties with long-term contracts. If one lived within one's image and did nothing to disappoint the behavioural expectations that image induced in the public, it was possible to have the proverbial 'swell time'. If, on the other hand, one rebelled against type, against formula, the studios could leave one high but dry, with few friends and fast-diminishing fortune.

Previously, success in American industry had been a unitary quality. The reward had been wealth, fame, and power. Those who succeeded received money, but fame went to the performers and power to the producers. The bosses, who felt they were the real risk-takers, looked on as their employees enjoyed the public acclaim and affection. The performers, who regarded this adulation as testimony to their importance, found that they lacked sufficient power to plan their futures. Rudolph Valentino had lamented, just before his death, 'A man should control his life. Mine is controlling me. I don't like it.' This tension reached a new intensity when Monroe became a star.

Andrew Tudor (1974: 80), considering this star–audience relationship, outlines four aspects that may contribute to the connection. 'Emotional affinity' is very common, occurring when the audience feels a loose attachment to a particular protagonist or performer. 'Self-identification' arises when involvement reaches a point at which the spectator 'fuses perspectives' with the star. 'Imitation' denotes a condition wherein the star is seen as seen as acting as some sort of model for the audience. 'Projection' is an extreme version of this condition, wherein the star-struck live their lives according to their knowledge of their favoured star. Although Tudor's taxonomy is helpful in explaining what audiences may do with the star images they are offered, it does not tell us why the offered images take the form that they do. Why *Monroe?*

Alberoni (1962: 93) argues that 'The star system . . . never creates the star, but it proposes the candidate for "election", and helps to retain the favour of the "electors".' Long before 'Superman came to the Supermarket', stars campaigned for their own office. The role and performance of a star in a movie were taken as revealing the personality of the celebrity (which then was corroborated by the stories in the magazines and on the newsreels). What was seldom seen and seldom stated by Hollywood or the stars was that this 'personality' was itself a careful construction and expressed only through the flow of films, stories, publicity and gossip. On odd occasions, the spectator might actually confront a star and say, 'Who do *you* think you are?' Fame homogenizes life, and creative impoverishment becomes a constant threat as the person comes to see her or his 'personality'. Cary Grant (Archie Leach), Tony Curtis (Bernie Schwartz), Doris Day (Doris von Kappelhoff), Marilyn Monroe (Norma Jeane Baker): the stars literally made their names, their names made their fortunes, and their fortunes made their fate – there was no going back. Class distinctions were replaced by cash distinctions in an America where the famous are encouraged to turn almost anything into capital. Technology abhors a secret, and the stars make self-interest legendary in the culture of confession. Each star is a reminder of the conspicuous waste of talent where, exploited and exhausted by the voracious media, careers are seen to be made and then fade.

Marilyn Monroe began her career when Hollywood was resorting to an almost Keynesian policy of spending its way out of trouble, with lavish musicals and glossy comedies. Her career ended with the studios near to bankruptcy. Monroe's image has to be situated in the pattern of ideas about morality and sexuality that characterized the fifties in America and can here be indicated by such instances as the flow of Freudian ideas in the post-war West (registered particularly in the Hollywood melodrama), the Kinsey report (1953), Betty Friedan's *The Feminine Mystique* (1963), rebel stars such as Marlon Brando, James Dean and Elvis Presley, the relaxation of cinema censorship in the face of competition from television, and so forth. These instances must themselves be placed in relation to the other levels of social formation, such as actual social and sexual relations and relative economic situations of women and men. Marilyn Monroe's composition of sexuality and innocence is part of that pattern, but one can also see her distinctive attraction for her audience as being the apparent condensation of all that within her. Thus, Monroe seemed to 'be' the very tensions that inflected the ideological life of fifties America.

Monroe was never a symbol of a sexual stereotype in the simplistic sense referred to by Sichtermann (1986); the very point about Monroe was the tension, her extraordinariness. Sexuality, as Foucault (1982) has stressed, is itself indicative of the process by which power relations are arranged and enforced. The questions we ask about 'sexuality', the fact that we deem it significant, reveal a more wide-ranging tissue of presuppositions. As Foucault's work shows, the construction of sexuality in discourse represents a further development of the movement in society towards the positioning of the body as an object of surveillance and control. Considerations on gender identity and sexual difference combine a range of notions centering on biological sex, social gender, sexual identification and sexual object choice. The assimilation of these in constructs of gender identity is an established process whose effect has been to establish a heterogenous and determinate set of biological, physical, social, psychological and psychic constructs as a unitary, fixed and unproblematic attribute of human subjectivity. In ideology, gender identity lies at the heart of human subjectivity. Gender is what crucially defines us; we are encouraged to perceive ourselves in terms of our sexuality, which is thus interpreted as the core of the self. Yet what is sexual in one context may not be so in another: an experience becomes sexual by the application of socially learned meanings.

The conventional image of Monroe is founded upon the kinds of male fantasies noted by Klaus Theweleit (1987). His study singles out two man-made archetypes: the 'White Woman' serves man's needs and withdraws into the home; the 'Red Woman' disturbs the man's composure, unsettling his 'masculine' image of self-control and moral strength. It is not simply that these feelings disturb the sense of masculine identity, but that the denial of these tender and erotic feelings establishes the very sense of male identity. Masculinity has to be constantly reconstituted in the continuous denial of vulnerability and 'feminine qualities'. As long as the sense of masculinity is built upon the systematic denial of 'feminine' qualities, men are left in a continuous and painful struggle with themselves, in constant anxiety and fear of the revelations of their natures. They believe they can control these fears within themselves, the fantasies about themselves, but they do so by projecting them on to women. Thus, as men learn to deny their emotionality, need and dependency, these parts of themselves do not go away, but rather find disguised forms within which to assert themselves. Sex, when admitted, is seen as a commodity, freeing it from concern with commitment, vulnerability, and caring in personal relations.

The man's description of woman vacillates between intense interest and cool indifference, aggressiveness and veneration. He yearns to possess her, yet when he does feel he can possess her he treats her with contempt. Fifties movie moguls demanded the stereotype of a blonde, the dream (so they said) of returning soldiers and of *Men Only,* something Michelangelo might have carved out of candy. The idealized movie 'Marilyn' was seen as sensuous, loveable and passionate, but, at the same time, scrupulously chaste. Monroe described those men who criticized her 'lewd' love affairs as 'White-masking' themselves: purifying themselves by projecting their 'Red' fantasies onto her. Wilde suggested that 'Men always want to be a woman's first love.' It is said that Monroe's former husband Joe DiMaggio began to resent her when she appeared so 'available': as is so often the case with men, he seemed to distrust what he could possess, and worship what eluded him (after their divorce he became her devoted admirer again). The mixture of desire and fear in the male fantasy Red Woman is certainly evident in Norman Mailer's *Marilyn,* where he describes her as 'a queen of a castrator'.

Blondeness is the ultimate sign of whiteness: the blonde woman is offered as the prized possession of the white man, the most desired of women, the most 'feminine' of women. In *Bus Stop* (1956), the man, looking for 'an angel', sees the Monroe character and exclaims: 'Look at her gleaming there so pale and white!' In *Niagara* (1952), Monroe made a rare appearance as a Red Woman, a woman who acts: she ends up killed by a man for her infidelity. In *Gentlemen Prefer Blondes* (1953), we find the now 'natural' image of 'White Marilyn', a woman who is usually acted upon: she ends up worshipped by men. The Red and White archetypes are most clearly captured in the 'blowing dress' scene used in *The Seven Year Itch* (1955): the Monroe figure is standing over a subway grating, away from any other human being, swaying with pleasure as the cool breeze blows about her legs, sending her skirt waving around her waist, her white dress on display for our gaze. In 1984, the same pose is found in *The Woman in Red:* Kelly Le Brock takes Monroe's place, her eyes open and staring at you, her red dress blowing above her waist to reveal her red panties, her dark hair stroking her bare shoulders. The White Marilyn is purely for inspection, not for touching, meant for no man but Everyman; the Red Kelly is obtainable, down from the pedestal into your arms. Monroe's image was marketed as playful and elegant and everyone's Marilyn, thus alienating no one. She was like the movies: we can all see her, *there,* but no one can keep her.

Monroe was certainly subjected to the complete range of technical and ideological strategies whereby the Hollywood of the fifties constructed the ideal-typical fifties 'sex symbol': narratives paced in order to allow time for the study of the image in between incidents; the costumes and contexts that exhibited the body to greatest effect; and the various camera techniques and lighting that worked to present the woman as an object to be observed. Unless critics believe Monroe possessed 'magical' qualities, it is not clear why the tricks of tinsel town made Monroe into a star when so many other ambitious models (some of whom were more 'typical' as embodiments of conventional ideas of 'beauty') were tried, tested and rejected. Audiences did not uncritically accept all that was presented to them. The movies helped to legitimize voyeurism on a mass scale, awarding the activity the safety of darkness. In a period experiencing quite considerable changes in lifestyles, the movie medium made a decisive separation of actuality and day-dreaming – this in turn undermining the society's dependence on the bonds between love, marriage, sexuality, and family.

The 1950s began with almost all of the World War II leaders still in place and a new war, the Korean, to fuel an economic boom in the USA. There the 'Affluent Society' was christened by Galbraith and took off in a dazzle of chromium tailfins, a quadrupling of so-called self-made millionaires, consumer advertising, built-in obsolescence and conspicuous waste. The 'ads' were dominated by the 'happy housebound housewife', raising more children than before, with an 'organization man in a grey flannel suit' by her side to supply an elaborately accessorized home. The 'materialist society' was not particularly attractive to everyone. The problem with this materialism was that it was really rather dull. There were no brave causes left, as Jimmy Porter was to say. East and West were frozen in postures of the Cold War. The young found Eisenhower's America achingly dull. J. D. Salinger's solitary rebel in *The Catcher in the Rye* (1951), Holden Caulfield, won a huge following with his scorn for his 'phony' elders. The first leather-jacketed motor-cycle gangs of rebels were personified by Marlon Brando in *The Wild One* (1954), a performance full of nervous mannerisms and mumbled lines. In 1955, a fully motorized teenager gained simultaneous fame and death when James Dean was smashed into a legend of steel and glass as a 'rebel without a cause'. Men could suddenly be seen on screen to be vulnerable and even weak – a trend which brought new rules for the 'insecure' personalities of homosexuals such as Dean and Montgomery Clift. Joe and John Kennedy, according to Collier and Horowitz (1984: 176), believed that 'Hollywood, with its ability to manufacture status overnight, would provide the aristocracy of the coming era. [People] found Jack fascinated not only by the tinsel and glitter but by the way that sex appeal, even more than sex itself, became power; by the way ordinary people came to inhabit the extraordinary celluloid identities created for them.'

Television networking began in 1948, rapidly coming to undermine Hollywood's prosperity. At the same time, a consent decree in a long-fought anti-trust suit forced the studios to divest themselves of their theatre chains, and the practice of block-booking was also abandoned. The studios tried to give the public what it could not get from the small screen: grandeur, spectacle, colour, and overwhelming personal involvement. An answer was sought in technological innovation: giant triple-screen Cinerama; smaller, cheaper one-camera versions of the same idea, like CinemaScope and VistaVision; finally, 3D and 'a lion in your lap'. Once the novelty had worn off (which was very soon), the problem remained what to put *on* the movie screen. What people wanted, it seemed, were stars. Yet the classic stars who had survived from the pre-war years were now maturing towards 'character' roles, and few had arisen during the forties to replace them. As the ageing silent star Norma Desmond cried out in *Sunset Boulevard* (1950), 'They don't make faces like that anymore!' Indeed, it had not been fully understood how much the star system had to do with the whole back-up structure of the Hollywood studio, managing, monitoring and manoeuvering every aspect of the contract player's professional life. If Marilyn Monroe was the last superstar to be shaped by the old mould, it was not so much because the raw material was necessarily lacking as because the breakdown of the studio system meant that mould also was broken.

What the new trends in the fifties meant was that there was no longer one vast, homogeneous movie-going public that went ritually and regularly to see movies outside their own homes. Post-war America offered a greater diversity of leisure pursuits than ever before, and consumers could afford to be more selective than previously. People now went to a particular movie rather than 'to the movies'. There were many different publics with many different interests. The emergence of a star who could draw a large number of these publics together seemed increasingly remote, but increasingly attractive to an insecure industry. The need for many productions, featuring many performers, declined during the decade and the institution of the long-term contract virtually disappeared. For those well-established stars, for whom there remained a demonstrable demand, the rewards were richer than ever. They could now insist upon more for a single film than they might have made for several years on a contract. As freelancers, they now had many highly agreeable options: the stars, although fewer, could shine ever brighter.

The general meaning of the myth of success is that American society is sufficiently open for anyone to get to the top, regardless of rank. David Boorstin (1962: 162) observes, 'The film-star legend of the accidentally discovered soda fountain girl who was quickly elevated to stardom soon took its place alongside the log-cabin-to-Whitehouse legend as a leitmotif of American democratic folk-lore.' 'Meet Marilyn Monroe,' trumpeted the 'fanzines', 'the star who came from nowhere.' During the Renaissance, fame was the reward for manifest deeds. Fame signified a social consensus about what was worthwhile; hence the traditional pairing of *fama* and *virtu*, 'prowess' or 'accomplishment'. The 'celebrity', as Boorstin noted, is famous for being famous: hence the aura of glorious gratuitousness and disposability. The myth of success is rooted in the conviction that the class system does not apply to America. This myth, developed in the star system, attempts to orchestrate several contradictory elements: that ordinariness is the hallmark of the star; that the system rewards talent; that luck typifies the career of the star; and that effort and efficiency are important for longevity. The career of Marilyn Monroe posed her chroniclers considerable problems: stressing her naivety and her 'patrons' left writers struggling to explain her struggle for independence and 'serious' roles; stressing her unpredictability and her determination made her personal problems difficult to explain. Monroe's struggle for autonomy was also a struggle against the traditional myth of stardom.

The struggle to situate the star has often resulted in still further mysticism. Stars, at their brightest incarnation, can come to be seen as of a different order of being, a different ontological category: Monroe became Mailer's 'sweet angel of sex', the biographers' 'tragic Venus', the studio's 'love goddess'. The star image, we can now say, is a construction composed of media texts that can be classed together as 'promotion, publicity, films and commentaries' (Dyer 1982). Promotion involves material produced especially for the manufacture of a particular image for a particular star: it includes studio and press releases, fan club reports (regulated by the studio), fashion photos, and public appearances. Less 'deliberate' in appearance is publicity – 'what the press finds out' – teased out in gossip and interviews. The public may here read tensions between the star-as-person and the star's image, tensions which at a higher level become themselves integral to the image (e.g. Monroe's campaign to transcend the 'dumb blonde' role). Films were often built around star images, serving as a 'vehicle' for a popular performer. Across Monroe's star vehicle one can discern certain 'family resemblances' – of iconography (how she is dressed, made up, and coiffed; distinctive gestures and settings); visual style (how she is lit, photographed, placed within the frame); musical motifs (the type of songs given to her; the background music that accompanies her appearance); and structure (her role in the plot, her function in the film's symbolic pattern) (see also chapter 4). 'Commentaries' include what is said or written about the star in terms of appreciation or interpretation by critics and writers; it involves contemporary and posthumous studies, and is found in film reviews, critical columns, scholarly books, and television and radio profiles. Monroe, in particular, was acutely conscious of the effect of commentary as a mediator between the audience and the actor; she cultivated several associations with columnists, bypassing her publicists and giving the information when she wanted and to whom she wanted. Early in her career she remarked:

> I might never see that article and it might be okayed by somebody in the studio. This is wrong because when I was a little girl I read signed stories in fan magazines and I believed every word the stars said in them. Then I'd try to model my life after the lives of the stars I read about. If I'm going to have that kind of influence, I want to be sure it's because of something I've actually read or written. (In Taylor 1983: 97)

The star's image is a complex totality, containing a chronological dimension. In order to grasp the temporality of this image we can refer to the notion of a *textured polysemy* (see Dyer, 1982). A star image is polysemic – it signifies several meanings and effects. This polysemy is textured: the several elements of signification may be mutually reinforcing (e.g. John Wayne as 'the all-American guy', the independent adventurer), or these elements may cause friction (as with Monroe's later image, characterised by her efforts to negotiate the difference between these elements). Images incorporate a temporal dimension: a change of roles may affect and inflect the image, as will a change in the person. Watching a star involves the use of hermeneutics: no 'reading' can be done purely in 'close-up', for each instance is impinged on by past instances and past interpretations. The star's personality in a movie is built up, negotiated by movie-makers and audience, across the whole movie. The audience arrives to watch a well-known star, armed with certain preconceptions about the performer derived from knowledge of past appearances, generic conventions, advertising campaigns, and critical reviews. It may be that the signals a star 'gives off' (Goffman 1971) are sometimes not deliberate, and yet constitute a major part of the star's image. As I note in Chapter 2, Monroe's parted lips give the signal 'yielding sexuality', but the quivering upper lip (in fact attempting to hide her high gum line) may also be read as giving the signal 'vulnerability'.

Marilyn Monroe came to exude an aura that affected audiences on an unprecedented scale. She was not the first struggling starlet to strive to symbolize something other or more than sex. The significant difference was that a melodrama, much larger, richer, yet apparently more intimate than anything she ever appeared in on the screen, could be constructed in real life around this ambition. The mass media was powerful enough to transmit any image around the world; the medium of the movies was anxious enough to exploit any image that could see-off the challenge of television; Monroe, when she began, was keen to escape from a wretched childhood and forge a brighter future. Eventually, the story would carry sub-themes of social and cultural criticism more potent than anything in the history of the star system, and it was assigned a firmly tragic denouement which has assured it a resonance that, if anything, seems to increase with the passage of time.

For Mailer (1973), the turning point came when Marilyn Monroe, her career beginning to grow, was faced with the threat of blackmail and blacklisting when an old calendar featuring her nude portrait was discovered. Monroe resisted her studio's advice and determined to make a public admission: yes, she had posed nude as 'Miss Golden Dreams' in order to get through a particularly lean time in her starlet days. Her confession brought a wave of public sympathy. Confession continued to play a key part in Monroe's success: her soft, shy, breathless delivery had the quality of an intimate, whispered expression, and her major press interviews were among the most strikingly direct of any celebrity this century. Monroe had tended to make her early life seem even more dramatic than it may have been; by 'novelising' her autobiography, imparting to it still greater pathos, Monroe linked her myth to the pre-existing myth of the victimised starlet, struggling to retain respectability in the exploitative environment of the movie business. A death-mask is said to display either a happy or sad expression, depending on from which side one looks, inside or outside; Monroe's image, seen on the screen and through her life-story, resembled this dialectic. As Mailer (1973: 94–5) comments:

> She . . . has burst out of all the standard frames of reference for publicity. . . Our heroine has been converted from some half-clear piece of cheesecake on the hazy screen of American newspapers (where focus always shifts) to another kind of embodiment altogether, an intimate, as real as one's parents, one's family, one's enemies, sweethearts and friends. She is now part of that core of psychological substance out of which one concocts one's life judgements. . . Marilyn had become a protagonist in the great American soap opera. Life is happy for her one hour, tragic the next; she can now appear innocent or selfish, wronged or wrongdoer – it no longer matters. She has broken through the great barrier of publicity – overblown attention – and is now *interesting:* she is a character out there in the national life, alive, expected, even encouraged, to change each week. The spirit of soap opera, like the spirit of American optimism, is renewal! God gave us a new role each week to watch, but a role that fits the old one.

Marilyn Monroe took her role as a 'dumb blonde' and played it for all and more than it was worth. The result represented the best that she could do with such a role, and people seemed to sense that she was not cheating or talking down to them through it, even though it read easily and did not tax as it touched. Monroe did not entirely understand the role of her role-playing, the space between the fiction and felt being. She was fascinated and frustrated by her multiple selves. The movie bosses failed to understand their most significant star: they typed her; they maligned her when she faltered and mocked her when she succeeded; they ignored her aspirations, but they would not let her go, for she was one of the few remaining stars worth holding on to. For movie-makers' whose lives began and ended with movies, Monroe's insecurities were irritants; for a public whose world was more than the movies, these very insecurities made Monroe seem more interesting. Foolishly, the studios kept underpaying her and casting her as the dumb blonde; Monroe came to satirise sex while still embodying it. She went to the Actors Studio, she employed a drama coach, she worked to improve her abilities in every direction. She married a legendary athlete, then later had an affair with a legendary politician. Her life was contorted and cut short by a career which was at the centre of a new social contract between politics and popular culture.

Cultural criticism refuses to entertain any simple notion of people as 'essences' existing outside of ideology. Consequently, the conventional bourgeois idea of the individual (with the associated sense of creativity, continuity, and competitiveness) has been sent into crisis. This sense of crisis is, I would suggest, central to the star phenomenon. How stars address, embody, or condense this crisis may well be predominantly in terms of reaffirming the reality of people as subjects over against ideology and history, or else in terms of revealing precisely the uncertainty concerning the definition of what a person really is. Marilyn Monroe, by virtue of her central position in celebrity culture, embodies this tension most acutely. Whether affirming or exposing, or moving between the two, Monroe articulates this crisis through the cultural and historical specificities of class, gender, race, and sexuality. Yet Monroe functions at the more general level – itself culturally and historically specific – of defining what a person is. A star like Marilyn Monroe, in revealing such a problematic, may be not merely of ideology but also *about* it. In so far as stars work progressively, they do so through audience identification with them. This identification may function either by effecting an affirmation of an alternative attitude by audience members to their own lives, or by elaborating an image of an escape from these life situations through models that suggest alternative modes of using or transforming them.

Whilst the principal aim of this book is the analysis and the demystification of the myth of Marilyn Monroe, it is not a denial of *her* – the presence and play of pleasure must perforce be admitted and accepted. Critical analysis should not be a conceptual neutron bomb, destroying living emotions and leaving structures intact. The text under analysis acquires its force and intensity from the way it is experienced, and that ideology inflects and informs the experiential and affective as much as the cognitive. My own fantasies and feelings are a necessary aspect of my analysis. It is of some relevance to this study and its reception that I experience a degree of pleasure and pathos in contemplating the myth we know as 'Marilyn Monroe'. At certain moments, with certain quotations, in certain images, I feel *moved* by the object of my study. The mythologist has to cope with the problem that in criticising something one is in some sense in complicity with it, for one articulates what goes without saying, inscribing mythical meaning. Writing a biography of Marilyn Monroe cannot erase previous biographies; making a montage of her image does not cut out involvement in iconography.

It is ironic that an orphan came to embody the anxieties of an era. Modern society suffers from the loss of community, neighbourhood and public life itself; celebrity fantasies have come to cover the cracks in the authentic public realm. No one grasped the gratuitous allure of celebrity with greater acuity than the self-styled dumb blonde, Andy Warhol. With Warhol the culture of packaging produced its paradigmatic artist. He extracted repetition from mass culture, making a product out of the process Walter Benjamin (1935) had noted thirty years before. 'I want to be a machine,' Warhol announced. He celebrated the peculiarly inert sameness of the mass produced: an infinite series of identical objects – Campbells soup cans, Coke bottles, dollar bills, and the portrait of Marilyn Monroe, silkscreened over and over again. Warhol's work addressed the fact that advertising promises that the same pap with different labels will give you special, unrepeatable gratifications. His montage of Monroe presaged the future decades wherein the unmistakeable image recurs like a rush of blood through every media channel: in books, in movies, on the cover of *Sergeant Pepper,* on thousands of T-shirts, prints and postcards, on millions of bedroom walls.

'Like Marilyn Monroe' is a directional arrow implanted on modernity. It rapidly overcame any obligation we might have felt to the facts of the life, to whether it was decent to wonder whether she was good as a performer, or whether we liked or dislike her. Legend does not trade in analysis or approval; it does require an actual nonentity, a suggestion of life, a subjunctive pervasiveness. The legend of Marilyn Monroe feeds on a film image which flits from low-comedy to heart-touching pathos, an image which proved the potency of popular cultural forms. An associate of the Kennedys is quoted in Collier and Horowitz (1984: 176) as saying: 'Charisma wasn't a catchword yet, but Jack was very interested in that binding magnetism these screen personalities had. What exactly was *it?* How did you go about acquiring it? Did it have an impact on your private life? How did you make it work for you? He couldn't let the subject go.'

For ambitious individuals in the 'Moronic Inferno', modern culture, the mechanism behind stardom was the key to unlock the present. 'Death,' it has been said, 'is an acquired trait' (Woody Allen 1980: 13). Certainly in death we have the image exactly where *we* want it. Those elements in the biography that in life discomfited fantasy now wither and eventually disappear, while the fantasies that we wished to impose upon the image can now be freely grafted on the unresisting ghost with no inconvenient facts to interrupt our reveries. A memory serving to efface remembrance: 'Marilyn' of memory, memory of 'Marilyn', a recurrent exchange, an acceleration to win time, win time over, disguise it, deny it. Death to the person is an airbrush for the image, imparting a satisfying glow to its subject. The myth of Monroe burns on and into the narrative repertoire as a symbolic representation of the 'American tragedy'. Who do we think she is, Marilyn Monroe? 'Everybody is always tugging at you. They'd all like sort of a chunk of you. They kind of like take pieces out of you. I don't think they realize it, but it's like "rrr do this, rrr do that. . ." but you do want to stay intact – intact and on two feet' (Monroe, in Meryman 1962).

Monroe said of her interviews: 'The questions often tell more about
the interviewer than the answers do about me.'
Photograph reproduced by permission of the BBC Hulton Picture Library.

2

Of Writers and their Inelegance

The truth is I've never fooled anyone. I've let men sometimes fool themselves. Men sometimes didn't bother to find out who and what I was. Instead they would invent a character for me. I wouldn't argue with them. They were obviously loving somebody I wasn't. When they found this out, they would blame me for disillusioning them – and fooling them.

Marilyn Monroe

If, to use a metaphor, one views the growing work as a funeral pyre, its commentator can be likened to a chemist, its critic to an alchemist ... [T]he critic inquires after the truth content whose living flame continues to burn over the heavy logs of the past and the light ashes of life gone by.

Walter Benjamin

'Those who know me better,' said Monroe, 'know better.' Monroe's biographers are boastful of having 'captured' her whilst wary of any suggestion of bias or conjecture. Narrative, they would say, tells itself. In opposition to the 'Recording Angel' is the realization that there is no 'natural' narrative and no 'objective' voice. Sartre (1976: 25–6), in his *La nausée*, follows his character Roquentin as he labours to delve ever deeper into old documents in search of 'the true Marquis de Rollebon'. To Roquentin's dismay, the sum total of all the facts and views do not add up to a definitive picture. How, he wonders, do historians achieve coherency?

How do they do it? Is it that I am more scrupulous or less intelligent? In any case, put like that, the question leaves me completely cold. At

bottom, what am I looking for? I don't know. For a long time,
Rollebon the man has interested me more than the book to be written.
But now, the man . . . the man is beginning to bore me. It is the book
to which I am growing attached.

More has been written about Marilyn Monroe, both during her
lifetime and since, than about any other figure in film hagiography.
Successive writers in successive studies have drawn out the image and
sketched in their ideal, told us all about 'Marilyn', that gentlemen prefer
'Marilyn', that 'Marilyn' made love, that 'Marilyn' was one of the misfits.
Images are invoked: 'Marilyn' in and out of control, in and out of bed, in
and out of love, in and out of luck – Exhibit A and Exhibit B, you are the
jury and the inquest has begun. Of the more than forty biographies
published, one finds a wide range of treatments and tastes, ranging from
Zolotow (1961) and Guiles (1971 and 1985) – both containing anecdotes
from Monroe's colleagues and critics – to the sober 'investigative'
approach of Summers (1985) and the 'factoid' flourishes of Mailer (1973
and 1980). All of these individual texts, I would suggest, can be most
rewardingly read as answers to a preconceived set of questions –
questions that are thoroughly permeated by social, cultural and political
presuppositions. Books on Marilyn Monroe are not so much about her as
about the authors – their values, their interests, their fantasies. Interpretation reveals more about the author as, suddenly, the lady vanishes.

When Grosset and Dunlap asked Norman Mailer to write a 25,000-
word preface to a picture book on Marilyn Monroe, the author was only
moderately interested. Motivated by financial pressures, Mailer began
reading Zolotow and Guiles, and found himself fascinated by the subject.
The 'preface', he decided, must become a *bona fide* biography. He told
Time magazine (16 July 1973): 'When I read the other biographies of
Marilyn, I said to myself, "I've found her; I know who I want to write
about".' As he worked, the piece grew first to 40,000, then 50,000, then
105,000 words. The book's original single-column design was changed to
a double-column format as the prose swallowed up the images. Mailer's
Marilyn (1973), which he would call 'a novel ready to play by the rules of
biography', was in many ways a love letter to the myth she left behind. 'In
her ambition, so Faustian, and in her ignorance of culture's dimensions, in
her liberation and her tyrannical desires, her noble democratic longings
intimately contradicted by the widening pool of her narcissism . . . we can
see the magnified mirror of ourselves, our exaggerated and now all but
defeated generation' (p. 17).

The significance of Mailer's treatment of the Monroe myth will be
given greater prominence in a later chapter, but at present his work helps

us to grasp something of the biographer's mind. Mailer admitted that his highly prejudiced attack on the role of Monroe's third husband, playwright Arthur Miller, was motivated out of jealousy (see also Chapter 6). Pauline Kael (1973) remarked at the time: 'Miller and Mailer try for the same things: he's catching Miller's hand in the gentile cookie jar.' In 1980, after the publication of his second book on Monroe, *Of Women and Their Elegance*, Mailer confessed: 'I always thought that if I had been a woman, then I would have been a little bit like Marilyn Monroe.' This may explain why his 'Marilyn' often sounds a little bit like Norman Mailer. In fact, in *Of Women and Their Elegance*, Mailer tried to enter her mind and assume her voice as if it were her autobiography.

Mailer tells me that 'Marilyn Monroe has always interested me because I find her nature so elusive, and modern psychology so inadequate for attempting to understand her.' Mailer's 'identification' with his subject is by no means unusual amongst Monroe's biographers. In fact, one may read one of two possible morals into the panoply of plots: either 'Dear Reader, I would have saved her' or 'Dear Reader, I would not have tried.' It is a common feature in biographies of Monroe – nearly all of which were written by men – to find the 'Miller' figure (the writer, the intellectual) 'cut around' and replaced, in spirit, by the current author. Such a movement is indicative of the more general historiographical process wherein certain facts are animated by their placement in a certain dramatic plot.

My intention in this chapter is primarily analytical: to discuss the professed aims of Monroe's important biographers, to elaborate the methodological strategies best suited to evaluate what in fact these writers do achieve, and to consider the very nature of the 'biography' itself. So much of what is active and ambitious in the biography is precisely what is overtly obscured by authorial pretensions to 'transparency'. Rather than being a window to a soul, the biographer's text is more of a distorting lens. Monroe is not an easily understood figure, and any single approach will experience a problem akin to the 'sheep shop' in *Through the Looking Glass* (1979: 68–9) where all shelves seemed full when viewed as a group but empty when seen in isolation. 'Things flow about so here!' Fascinated, the analyst will follow Alice: 'I'll follow it up to the very top shelf of all. It'll puzzle it to go through the ceiling, I expect!' The 'real Marilyn Monroe', I want to suggest, is a proper appreciation of her fictions, even if they are facts; or her facts, as long as one is not certain that they cannot serve as fictions.

Norma Jeane Baker (the spelling on Monroe's birth certificate shows 'Norma Jeane', and Monroe expressed a preference for the original spelling) was born on 1 June 1926, became one of the most famous people in the world, and died in mysterious circumstances on 4 August 1962. These two dates form quotation marks around a life which many biographers have found (according to the jacket notes of *Goddess*) 'as absorbing as any fiction'. Each writer has set out to strip the myth bare, to retrieve the body from the literary embodiment, to find the flesh and blood of Marilyn Monroe. Anthony Summers (1985: xiii) explains that, to his astonishment, 'not a single author had attempted an in-depth examination of the actress's personal life, coupled with serious inquiries into her death and the alleged connection with the Kennedys. My publishers, with considerable courage, then invested in my belief that there was a fresh and stimulating book to be written.' Summers' claim on his readers' attention rests on his reputation as an 'investigative journalist' rather than on the uniqueness of his approach: he has, we are told, steadier hands and sharper eyesight. His publishers make the aim eminently clear: to remove the clothing from Clio, to make History reveal its truths. *Goddess,* the jacket notes tell us, will disclose Monroe's myth; such phrases as 'reveals', 'casts light on', 'lays bare', 'exposes', and 'uncovers', emphasize the implicit ideological presumptions at work (see Heidegger 1971). 'With *Goddess,* it is announced, 'all previous books on Marilyn Monroe become redundant.'

Is Summers the highly trained historical paint-stripper he sets out to be? Early on he asks, 'What was at the hidden center of the phenomenon that was Marilyn Monroe?' (p. 4). Although his work succeeds in asking the questions of his predecessors with greater rigour and empirical effort, nevertheless he is still essentially reiterating. Summers begins his book with the same quotation (*The Wings of the Dove* by Henry James) as W. J. Weatherby opened his 1976 *Conversations with Marilyn:* 'You're right about her not being easy to know. One *sees* her with intensity – sees her more than one sees almost anyone; but then one discovers that that isn't knowing her.' Thus Summers makes a symbolic *rapprochement* with his predecessors, and perhaps unconsciously admits his doubts about really 'finding Marilyn'.

From Zolotow (1961) onwards, the holy grail has been hunted with boundless energy. Hoyt (1967: 9) begins with one of the clearest expressions of this curious interest:

> On beginning this study of an American motion-picture star, the
> author had fewer preconceptions than most writers might be expected
> to have. I had never seen Marilyn Monroe in the flesh. I had never
> seen a Marilyn Monroe motion-picture... My interest in Marilyn was
> kindled by the manner of her death and the obvious shock it brought
> to so many different types of people, who seemed to care so much
> about this strange girl.

Typically, Hoyt admits that 'so much has been written . . . and so little truth has been revealed. Marilyn was a difficult subject to catch.' There is something notably romantic about this urge to 'pin' Monroe down, to 'dominate' the facts. One might think that the books of the 'Marilyn speaks!' variety escape from this ill-conceived ideal, but this is not the case. Taylor (1985: 9) declares that, 'I thought it was about time Marilyn Monroe spoke for herself.' This laudable thought is hindered by a structure which forces comments out of their context and into several rather debatable categories. Monroe's autobiography was given a crude gloss by her male ghostwriter Ben Hecht. Hecht's (1954) text, written as a Monroe autobiography, appeared in book form in 1974 as *My Story*. Whilst certain passages exhibit a sub-Dickensian style which must surely be Hecht's doing, there are places wherein Monroe's voice seems authentic. With these reservations, the material is invaluable for the biographer.

W. J. Weatherby (1976) presents, according to the jacket notes, a 'revelation, in her [viz. Monroe's] own words, of her own thoughts and feelings. It is probably the closest we shall ever come to knowing the real Marilyn, the persona behind the pin-up photograph, the tinsel glamour and the Hollywood publicity machine.' The author has told me that, 'I was disturbed by the way in which several pseudo-biographers had merely used [Monroe] and so I felt I might help to set the record straight as she was someone I liked very much.' Weatherby proposes to keep his own comments to a minimum, in order to 'let her speak for herself' (1976: 4). In fact, there follows the most curious example of the 'author-as-actor' in the narrative. Monroe is cast opposite 'Weatherby', a character redolent with the sharp, shrewd features of Raymond Chandler's hardboiled sleuth, Philip Marlowe. Indeed, 'Weatherby' reminds one of Woody Allen's *Play It Again, Sam* Bogart parody, moving the narrative onwards with *film noir* nous: 'I was determined not to become obsessed with Monroe, as so many journalists had' (p. 25). 'She stared at me as if she was about to speak, but I hurried on' (p. 42). '[N]ow she was trying to woo me, to win over yet another journalist, another interviewer. I had been through that experience before with movie stars and politicians, and she began to lose me' (p. 60). Allen's Bogart is an outrageous misogynist who claims that there 'is nothing a little bourbon and soda won't fix'. 'Weatherby' seems to sympathise: 'I had been invited for a drink because I had seemed not to be interested. If you showed lady luck the back of your hand, sometimes she came crawling' (p. 66). The text then begins to feature Monroe's conversations with Weatherby, and there follows some genuinely charming and often touching passages, eventually serving to undermine the earlier self-indulgence. His memoir is one of the most valuable and respectful ever written on Monroe. Weatherby's text represents a rare case of the 'subject' overcoming the vagaries of the form.

One would expect that, in the few commentaries on Monroe written by women authors, a less condescending account would be found; the reality is somewhat different. The problem faced by these women writers is whether Monroe should be an identification figure for women in general. Diana Trilling's 'The Death of Marilyn Monroe' (1963) sees Monroe's life and death as 'a tragedy of civilisation' – a 'universalist' perspective. Trilling claims that we desire to look at sex without illusions, yet to look at sex is just what 'civilization' cannot permit; Monroe mediated between this opposition by 'performing as if' she was presenting us with the reality of sex. Joan Mellen's *Marilyn Monroe* (1973) counters Trilling's argument, choosing to stress the negative value for women of Monroe's image. Mellen finds Monroe, the person, as a victim of Hollywood and patriarchal society, pitiable and vulnerable, but the films are wholly demeaning. Although Mellen's observations on the movies as 'vehicles' has some legitimacy, nowhere does she examine the movies *as movies,* and Monroe's multi-faceted performances are reduced to Mellen's fixated concern. Molly Haskell's analysis of women's representation in movies, *From Reverence to Rape* (1974), develops a mini-biographical study of Monroe which is more sympathetic than Mellen's. According to Haskell, 'throughout [Monroe's] career, she was giving more to idiotic parts than they called for – more feeling, more warmth, more anguish; and, as a result, her films have a richer tone than they deserve' (p. 256).

One of the most publicized of recent biographies, *Marilyn* (1986) by Gloria Steinem, is also one of the most disappointing. The book is dedicated to 'the real Marilyn', a Marilyn sounding rather like Gloria Steinem. A series of self-contained essays follow, with Steinem promising that 'major themes will be repeated from different viewpoints in several essays, so that a factual and emotional holograph of a real person will begin to emerge' (p. 2). Steinem claims that the public Marilyn Monroe carried 'the real Norma Jeane inside her': with sadly typical crudity, the 'Marilyn' dustjacket slips off to reveal an indelibly engraved 'Norma Jeane' on the card cover. According to Steinem, inside the sex symbol was an abandoned, lonely, neglected little girl. This 'theory' is presented in the language of pop psychology as though it were a startlingly innovative thought: the star as victim. By focusing so exclusively upon Monroe's vulnerability, Steinem has unconsciously given the legend a decidedly *anti-feminist* slant. Monroe actively participated in the creation of her legend, and to play this down is to make her less powerful and impressive than she really was. Steinem does not even maintain a consistent argument, alternating between sympathy and scepticism: in one chapter she writes, 'other than Marilyn's words to friends. . . there is little evidence of her dozen or so illegal abortions,' whilst in the conclusion she writes, 'As her twelve or thirteen abortions testify, she was able to conceive a child, but may have preferred to remain one.' Contrary to Steinem's case, there is no proof of any such 'abortions', and it is astonishing that such a 'sympathetic' writer should place so much trust in an array of dirt-digging memoirs and Hollywood hearsay. There was strength as well as vulnerability in Monroe, and to deny her any responsibility for her own fate is to do what the dominant culture did to her: to return her to the status of an irresponsible. Steinem has contributed so much of value to the feminist movement – this makes *Marilyn's* inconsistencies all the more unfortunate. Women are far from being 'passive victims', and Monroe is weakened and sentimentalized in Steinem's treatment. The appearance of a sensitive biography of Monroe by a feminist, showing her both as a person *and* as a woman, is still urgently needed.

Monroe said of the 'image' she was given by the press: 'I'm close, I can feel it, I can hear it, but it isn't really me' (in Taylor 1983: 27). The question of what or who we experience when we read a biography is a complex and criticial issue. Hayden White (1978) has used Northrop Frye's notion of 'emplotment' to explain, not the 'hermetic' character of the historical work, but its relation to a limited number of archetypal 'plots', such as tragedy and comedy. In most modern biographies, as Lowenthal (1961: 119) notes, success is based not on work but on luck. With Monroe, from Guiles onwards, the metaphor of a 'candle in the wind' has been particularly influential: she is a 'product' of her background, by virtue of a primitive Darwinian concept of social facts. Lowenthal adds that the child (e.g. Norma Jeane) is seen as 'a midget edition' of the adult; indeed Hoyt repeatedly refers to Norma Jeane as 'the girl who would be Marilyn.' Lowenthal concludes: 'people are not conceived as the responsible agents of their fate in all phases of their lives, but as the bearers of certain useful or not so useful character traits which are pasted on them like decorations or stigmas of shame' (1961: 125).

Anthony Summers begins his story of Monroe with her death – her life forming a lengthy 'flashback', haunted by the ghost of the first chapter, ending with a part entitled 'The Candle Burns Down'. Guiles' very title, *Norma Jeane,* bespeaks of the subsequent structure (the child meeting an inevitable, tragic fate). Hoyt's contents page indicates the chronological (and 'vertical') pattern of the narrative: 'An Empty House', 'The Bottom Rung', 'Waiting', and later, 'A New Marilyn', 'No Answer', and 'Tragic Innocent'. Mailer, in many ways the most 'honest' of biographers in his admission of his 'novelistic' approach, quite clearly illustrates the rhetorical, persuasive labour of Monroe's biographers (they may record, but – as Mae West would say – 'they're no angels').

Biography is essentially a demythologizing form. It functions to correct, restate, revise or reinterpret false or distorted accounts of the subject. Freud remarks to a prospective biographer (Freud 1970: 127): 'Anyone who writes a biography is committed to lies, concealments, hypocrisy, flattery and even to hiding his own lack of understanding, for biographical truth does not exist, and if it did we could not use it.' Freud's own conceit was to persuade his subjects to research their own biography. Generally, through fact and revision, biography strives to demythologize the individual; inevitably, this becomes an ironic effect, since readers replace old myths with new if they read biography uncritically. The representational aspect of a life, a picturing of the experiences of a single person, become elements of a universal type. In universalizing the narrative, drawing on archetypes and conventions, biography moves from the realm of history to that of myth. Different biographers will infer different intentions, discern different values: the more one speculates, the greater the tendency to slip into covert autobiography (witness Mailer). Whichever biographer we decide to trust, their opinion will again be a text that we will have to interpret. This leads logically to an infinite regress, stopped only by an act of will. That is, we arrive at the authentic meaning precisely when we *decide* we have arrived there: we *make* the meaning, we make much of the meaning, and we sometimes are *moved* by the text to make it in a particular way.

We can discount the chimera of a 'definitive' account of Marilyn Monroe. The biographer-as-magician, tapping the black box to signal the emergence of a walking, talking, living 'Marilyn', is still common but hardly credible. When we examine 'Monroe by Zolotow' or 'Monroe by Summers', we can indeed discern White's 'deep-structural' coherence. The 'generic story-type' does not necessarily conflict with the 'purely interpretative' level of the text, but it would be almost impossible to disentangle one from the other, without reducing the biographical text to a kind of anaemic algebra. Barthes (1981) stresses that historical discourse rests on 'enthymematic' reasoning, on rhetorical rather than demonstrative argument. With Monroe, the body is already in the library (a murder may have occurred), the biographer has pleaded for justice, and *you*, 'Dear Reader', are the jury.

The difference between the 'realistic' claims of the novelist and the historian would reside precisely in the fact that the novelist admits the level of signification: the novelist knows that the text depends upon such mental constructs as genre and plot, while the historian is reluctant to assume any such thing – hence the stir caused by Mailer's *Marilyn*, which seemed to be 'playing around' with a 'real person'. If we try to read the Monroe biographies *as if* they were fiction, we must surely recognize that the inappropriateness of doing so does not spring from any observable linguistic features of the text as such (compare 'writing a thriller' with 'writing about Marilyn Monroe'). It is simply that we have learnt how to distinguish between fiction and history as making different truth-claims for their individual descriptions (see Genette, 1982). The evidence that we have learnt this capacity can be located in our disposition to insert the signified 'it happened' behind each instance of the past tense in an historiographic context.

It will thus be enlightening to continue this chapter with an analysis of how Monroe's written life has come to acquire the 'shape' it has, with its dramatic highs and lows, its patina of 'tragic' destiny. The story, typically, would appear as a frown if charted point by point: beginning low (orphan), rising to a peak ('sex goddess'), and falling to its end (suicide or murder). The biography is a kind of 'pinball machine': the plot is planned; the supporting cast (Norma's mother, Jim Dougherty, Joe DiMaggio, Natash Lytess, Arthur Miller, the Strasbergs, the Kennedys) are positioned beforehand, often 'static' in terms of personality; finally, 'Marilyn' is fired into this narrative by the author, made to bounce off each figure until the game is completed. Although this may have the appearance of an unfairly harsh account of what Monroe's biographers 'do', it is quite accurate, I believe, in describing the mechanics of their narratives. In what follows, Monroe's life story will be charted chronologically in order to clearly chart the development of themes and images by writers both during and after her own lifetime.

Norma Jeane Baker was born 1 June 1926, in or near Los Angeles, under circumstances whose mysteries, after disturbing her childhood, would be seen to aggravate her mature anxieties. Of her father, little is known; it is likely that he was C. Stanley Gifford, an employee of Consolidated Film Industries. In any case, he was effectively absent from the beginning and, as a girl, Norma Jeane dreamed of a father who resembled Clark Gable. Of her mother, more is known. Gladys Mortensen, a divorcee, was a film-cutter at the processing laboratories of Consolidated Film Industries; she had been betrayed, abandoned, widowed and adjudged insane. She in turn abandoned Norma Jeane to a sequence of orphanages and foster homes, the child passing through twelve families in the process.

Contrary to common accounts, Monroe was not 'doomed' to psychiatric illness, but she was born at some risk. Her maternal grandmother, Della Monroe, died in an asylum, aged fifty-one, just over one year after Norma Jeane's birth. The cause of her death was officially given as heart disease, with 'manic-depressive psychosis' as a contributory factor. Monroe herself later recalled, and a number of biographers record, an incident when her grandmother was moved during an hysterical fit to attempt to smother her. (Monroe's account was repeated by Arthur Miller to Guiles; Miller alludes to the event in *After the Fall*, 1979.) Anthony Summers (1985), generally sceptical of the 'childhood stories', notes the absence of corroborating evidence. Della was, however, almost immediately committed to the asylum. There followed what no one would disagree about – an appalling childhood: ten foster homes, two years in the Los Angeles Orphans' Home, another foster home, and finally four years with the guardian appointed by county authorities. An influence which surely exacerbated the young Norma Jeane's insecurity was 'aunt' Ana Lower and her constant teaching of Christian Science. This religion, the one pushed on to Norma Jeane, emphasized the power of 'right thoughts' in allaying pain and suffering: one is ill due to one's sins, one must work to 'happify' one's existence.

The social environment of Los Angeles during Norma Jeane's childhood was undergoing almost constant significant changes. Industrially, the introduction of 'talking pictures' encouraged further expansion in the movie industry. Clara Bow, Norma Shearer, Colleen Moore, Greta Garbo and Alice White were popular women stars; among the leading men, Richard Dix, Richard Barthelmess, John Barrymore, Adolphe Menjou, and John Gilbert dominated the screen. Rudolph Valentino died the year of Norma Jeane's birth. Charlie Chaplin and Buster Keaton were both trying to adjust to sound movies. The studio system in Hollywood was becoming more powerful, importing talent from Britain, Germany and France, and producing long-term contracts of awesome complexity to secure the services of local talent. For the majority of Californians, the Depression struck deep, and social mobility was radically reduced. Norma Jeane's legal guardian, Grace McKee, married a man whose job necessitated moving East. For reasons both of finance and of convenience, a marriage was arranged between sixteen-year-old Norma Jeane and a young neighbour, James Dougherty. It was a marriage which Dougherty (1976) later wrote was relatively rewarding; his former wife recalled that the marriage brought her neither happiness nor pain, just an aimless silence.

Norma Jeane was left to learn to be a housewife while her husband worked at Lockheed. Eventually she began planning for dinner after breakfast; peas and carrots were regular constituents because of the colour combination. Dougherty joined the Merchant Marine in autumn 1943. When he was abroad, his wife took a job at Radio Plane, a plant producing aircraft used for target practice; she inspected parachutes and

sprayed fuselages. Private David Conover was an army photographer for
an armed service motion picture unit. His commanding officer was
Captain Ronald Reagan. Conover's mission at Radio Plane, was 'to take
morale-boosting shots of pretty girls' for *Yank* magazine. Conover (1981)
recalls that he was struck by the photogenic quality of Norma Jeane, and
persuaded her to pose for him.

It is usually at this moment that the 'childhood' section of Monroe's
story is left behind, together with many of the most troubling aspects of
her 'traumatic' upbringing. Monroe recalled a childhood assault:

> I was almost nine, and I lived with a family that rented a room to a
> man named Kimmel. He was a stern-looking man, and everybody
> respected him and called him Mr Kimmel. I was passing his room
> when his door opened and he said quietly, "Please come in here
> Norma. . ." He smiled at me and turned the key in the lock. "Now
> you can't get out", he said as if we were playing a game. I stood
> staring at him. I was frightened, but I didn't dare yell. . . When he put
> his arms around me I kicked and fought as hard as I could, but I
> didn't make any sound. He was stronger than I was and wouldn't let
> me go. He kept whispering to me to be a good girl. When he unlocked
> the door and let me out, I ran to tell my "aunt" what Mr Kimmel had
> done. "I want to tell you something" I stammered, "about Mr
> Kimmel. He- he-" (In Hecht 1954)

Monroe claimed that her foster parent told her, 'Don't you dare say
anything against Mr Kimmel. Mr Kimmel's a fine man. He's my star
boarder!' Kimmel later told Norma Jeane 'to go buy some ice cream'.

I would cite this incident as an example of the reading of the adult in the
child that Lowenthal mentioned. Monroe's biographers tend to imply
(and sometimes suggest) that she had either encouraged the man or been
unconcerned. Summers believes Monroe may well have greatly embell-
ished the story (if it occurred at all), and cites people who claim that in her
last year alive she 'was still rambling on interminably about the assault'.
Yet surely, if an eight-year-old child *was* sexually molested, she or he
would not be extraordinary in 'rambling on interminably' about the
experience. Such a phrase is more indicative of her listeners' gross
insensitivity.

Monroe is described as having a 'borderline' personality (see, especially, Summers 1985). The development of her kind of disorder is highly
compatible with a childhood history of sexual abuse. Significantly,
Norma Jeane was molested when eight years old; at this young age, *pace*
Summers' claims, a child rarely has sufficient knowledge to fabricate such
a story – unless she or he has really been exposed to some sexually explicit

encounter. Biographers from Zolotow to Mailer and Summers perpetuate the myths surrounding Monroe's childhood by their voyeuristic mode of presentation and their sexist interpretation of her life. Victims of childhood sexual abuse are often emotionally stunted at an early age. They put much psychic energy into dealing with the abuse, and the natural spirit towards maturity is threatened. Her sometime-lover, Fred Karger, told her that she cried too easily: 'That's because your mind isn't developed. Compared to your figure, it's embryonic' (in Summers 1985: 42). The result of such trauma is a child in a woman's body, portraying a certain naivety by tending to play the seductress in the long and painful process of trying to come to terms with her abuse, and with her perception of men. Monroe told W. J. Weatherby (1976: 144):

> There were times when I'd be with one of my husbands and I'd run into one of these Hollywood heels at a party and they'd paw me cheaply in front of everybody as if they were saying: *Oh, we had her.* I guess it's the classic situation of an ex-whore, though I was never a whore in that sense. I was never kept; I always kept myself. But there was a period when I responded too much to flattery and slept around too much, thinking it would help my career, though I always liked the guy at the time. They were always so full of self-confidence and I had none at all and they made me feel better. But you don't get self-confidence that way. You have to get it by earning respect. I've never given up on anyone who I thought respected me.

Although the biographers dramatically document that 'psychotic' characteristics were applicable to Monroe, they signally fail to understand their significance. Instead, the authors portray Monroe as a manipulative, conniving female with only her own interests at stake, and with a ruthless drive to succeed at any cost – a woman who 'cries too easily' and 'flirts too often'. The very qualities resulting from Monroe's childhood abuse were further exploited by Hollywood film-makers and publicists. She became a 'sex goddess', a child to be worshipped in a woman's body. Although there is a publicly recognized taboo on the sexual molestation of children, nevertheless, as a society, we mould women into children. Monroe thus held a special attraction for many men, as her clothes designer Billy Travilla so graphically describes: 'She was both a woman and a baby. . . A man wouldn't know whether to sit her on his knee and pet her, or put his arms around her and get her in the sack' (in Summers 1985: 81–2).

Norma Jeane had little time to reflect upon the problems of her enforced independence. Her modelling career began in earnest when Emmeline Snively, head of the Blue Book Model Agency, started handling her affairs. Snively found Norma Jeane in need of tuition for her new profession: according to Snively, her protégée's nose was too long, there was insufficient upper lip between the end of her nose and her mouth, and thus she was taught to 'draw down' her upper lip when she smiled. Years later, after stardom came, her lips would still quiver as she smiled, and the movement became part of her style. Jimmy Starr, a journalist, claims that Monroe 'invented' her distinctive walk to capture the camera's attention: 'She learned a trick of cutting a quarter of an inch off one heel so that when she walked, that little fanny would wiggle. It worked' (in Summers 1985: 44). Aside from the asininity of Starr's attitude, he is also quite mistaken. Monroe's 'walk' was a natural one, occasionally made into a self-parody. Snively first noted that Norma Jeane's knees locked, throwing her pelvis back and her chest forward in a gesture that was natural, yet as a pose it was less than graceful for a model. Norma Jeane's soft, husky voice was later highlighted when Edmund Golding, directing *We're Not Married* (1952), told her to take quick short breaths before speaking her lines. Snively also persuaded her to dye her hair blonde. The 'image' began to be worked upon. Norma Jeane's portrait began to appear on the covers of *US Camera, Peek, Laff* and *See*. Summers (1985) repeats the usual story of Howard Hughes, hospitalized after an accident, coming across Monroe's image in one of these magazines. However, Snively says that she invented the story, sending it to Hollywood columnists Hedda Hopper and Louella Parsons to stimulate movie studio interest (see Guiles 1971).

Norma Jeane was sent to see the head of casting at Twentieth Century-Fox, Ben Lyon. An enthusiastic Lyon began to help Norma Jeane: he advised her to call herself 'Marilyn' (after Marilyn Miller), and she chose 'Monroe' as it was her grandmother's surname. After a brief spell with Fox, under contract, she was dropped. As her financial situation became particularly dire, she accepted fifty dollars to pose anonymously nude on red velvet for photographer Tom Kelley. The pictures, produced as a calendar, brought Kelly nine-hundred dollars from a printer who sold it in quantity for a considerable sum all over America. When Monroe was later on the verge of stardom, her identity as 'Miss Golden Dreams' was revealed. Her witty handling of reporters' questions – 'Didn't you have anything on?' 'Just the radio' – and her explanation – 'Sure I posed. I was hungry' – actually improved her public image.

By 1947, divorced and without a contract, Monroe lived quietly at the Studio Club. She read voraciously, rarely went out, and laboured to 'improve' herself. She mispronounced words, and occasionally suffered from a stutter she had acquired in early childhood. She joined acting, dancing, and singing classes: 'I knew how third-rate I was. I could actually feel my lack of talent, as if it were cheap clothes I was wearing inside. But, my God, how I wanted to learn! To change, to improve! I didn't want anything else. Not men, not money, not love, but the ability to act.' (Monroe 1974: 54)

Monroe studied *De Humani Corporis Fabrica,* an account of the human anatomy by Andreas Vesalius; she marked it up in detail, and even at the end of her life she would still instruct young friends with an encyclopaedic knowledge of the human bone structure. She was one of the first joggers in Los Angeles, diligently working to improve her fitness. She studied Mabel Elsworth Todd's *The Thinking Body,* a work which attempted to situate the body within the continuum of human psychology and physiology.

After a brief spell with Columbia, Monroe was taken back into the fold of the Fox studios with a standard contract arranged by Ben Lyon and Fox executive producer Joseph M. Schenck. Guiles (1971) claims that 'abstract ideas' always remained outside Monroe's grasp. Russian emigré Schenck, says Guiles, was a good 'match', for his 'approach to life was tactile' (!). The 'tactile' method, aided by Lyon's encouragement, led to Monroe's role in John Huston's *The Asphalt Jungle* (1950), a good part in a fine thriller which brought her recognition and the offer of another role – as aspiring actress Miss Caswell, a 'graduate of the Copacabana School of Dramatic Art', in *All About Eve* (1950).

In February 1951, Monroe enrolled in courses in the adult extension division of the University of California at Los Angeles. She took courses in art appreciation and literature, focusing on the Renaissance. To her literature teacher, 'she could have been some girl who had just come from a convent'. Monroe was a quiet student who worked well, until a fellow student noticed her portrait in a film magazine and announced her identity. She was forced to withdraw after the classes became public knowledge. *All About Eve's* director, Joseph Mankiewicz, found Monroe reading Rilke's *Letters to a Young Poet.* Intrigued, Mankiewicz asked her how she came to select the book; she replied,

> I was never told what to read, and nobody ever gave me anything to read. You know – the way there are certain books that everybody reads while they're growing up? . . . So what I do is – nights when I've got nothing else to do I go to the Pickwick Bookstore on Hollywood Boulevard. And I just open books at random – or when I come to a page or a paragraph I like, I buy that book. So last night I bought this one. Is that wrong? (In Hoyt 1967: 86)

Mankiewicz told her no, it was not wrong, there was no better way to select books for reading.

Monroe's personal library – which included works by Freud, Marx, Locke, Tolstoy, Proust, and Stanislavsky – has been the object of much derision by her colleagues and her biographers. Hoyt (1967: 152) is typical in his condescending remark: 'Perhaps if she had started with something a little less taxing, such as *Treasure Island* or *Little Women,* Marilyn would have found the world a more pleasant place.' Jack Paar, a minor actor, who once appeared with Monroe in *Love Nest* (1951), went on to discuss what he saw as Monroe's 'pathetic' desire to impress. Recalling her interest in Proust, Parr (1983: 85) explains to us that the author was 'a rather exotic French author much in vogue among the intellectually pretentious of the time.' Having placed Proust as a fad of the fifties, the misogynism in Parr's account becomes more evident: 'I fear that beneath the facade of Marilyn Monroe there was only a frightened waitress in a diner.' Such interpretations abound in the literature on Monroe, and there is no evidence to support any of them.

The desire to improve her intellect and her acting was always regarded as 'eccentric' by studio heads, delighted by the commercial appeal of their 'dumb blonde'. In later years, her association with psychiatrist Ralph Greenson probably only exacerbated her self-doubt: Greenson was one of the strongest critics of Kleinian psychoanalysis, and thus a sceptic of Klein's arguments for women's creativity. Monroe fiercely reacted against the public image she was given: she seemed childlike and foolish, she said, but 'That's because of the parts I play. If I play a stupid girl and ask a stupid question I've got to follow it through. What am I supposed to do – look intelligent?' (in Taylor 1985: 120). Breaking out of this 'type' was not easy: Hollywood was impatient with 'independent' women, although male stars were allowed more control in their work. When Monroe played a psychotic child-minder in *Don't Bother to Knock* (1952), she was attacked by the critics. *Niagara* (1953), a camp melodrama directed tongue-in-cheek by Henry Hathaway, saw Monroe in a slightly more familiar role as a *femme fatatle*.

How to Marry a Millionaire and *Gentlemen Prefer Blondes* were released in 1953. In the latter, Monroe dances *à la* burlesque, her bumps and grinds pruriently denatured to satisfy a code which, forbidding nakedness, provides the raw material from which interested persons may labour independently upon their own fantasies. A positive sign from this period is the very evident rapport between Monroe and her co-star, Jane Russell. In her own autobiography (1986), Russell gives a warm account of Monroe's capacity for forming close friendships with other women – a capacity often ignored by her biographers. Russell recalls that Monroe enjoyed the company of women, delighting in the relaxed atmosphere and the sense of shared values; Monroe spent much of her time comparing notes with Russell on how to cope with the limited cultural interests of their respective athletic partners.

For Marilyn Monroe, a formula had been found for her movie work. Henceforth she was compelled to perform according to the formula for as long as the profits flowed. Monroe was now the leading box-office attraction at Fox. Upon marrying American baseball hero Joe DiMaggio, she was one of the most prominent international celebrities. In 1954, Monroe filmed the now-famous 'skirt' scene for *The Seven Year itch* on a New York street-corner, watched by a mass of people. DiMaggio, an extremely temperamental man, was horrified at his wife's exhibitionism. That night, screams were heard from the DiMaggio suite. Gladys Whitten, Monroe's hairdresser, met her in the morning. 'She said she had screamed and yelled for us,' Whitten recalls, 'Her husband had got very, very mad with her, and he beat her up a little bit. . . It was on her shoulders, but we covered it up, you know . . . a little make-up, and she went ahead and worked' (in Summers 1985: 103). Amy Greene, Monroe's New York friend, also saw her bruises: 'Her back was black and blue – I couldn't believe it. . . She didn't know what to say, and she wasn't a liar, so she just said, "Yes. . .".'

Monroe gave DiMaggio a gold medal for his watch chain, inscribed with a quotation from *The Little Prince* by Saint-Exupery: 'True love is visible not to the eyes, but to the heart, for eyes may be deceived.' A perplexed DiMaggio exclaimed, 'What the hell does *that* mean?' Monroe was cultivating tastes for Vivaldi and Bach, Whitman and Yeats. DiMaggio was still transfixed by television, a connoisseur of comic books. He warmed to talk of him at Toots Shor's club as a womanizer, yet he winced at any suggestion that his wife was in any way unfaithful. Roger Kahn's *Joe and Marilyn* (1987) is a manly account of the marriage, written by an author who knows much about men and baseball but apparently little about women and movies. There is no doubt concerning Kahn's identification with DiMaggio: Monroe, he decides, 'was infinitely self-centred . . . dead by her own hand because she wanted to die more than she wanted to live.' DiMaggio, Kahn says, 'wept the hard, hot tears of a man who does not like to cry' (p. 202). Biographers, cognizant of DiMaggio's lasting loyalty to the memory of his former wife, have usually produced either a saintly symbol of all-American manliness (which makes Monroe's divorce from him seem quite irrational), or a caricature of the 'dumb sportsman' (as in *Insignificance,* which makes Monroe's love for DiMaggio seem altogether irrational and masochistic). Regrettably, as each author contorts the story, DiMaggio is further stereotyped and subdued.

Throughout the fifties, a large public waited and watched as Monroe tried to have a baby. The headlines told of repeated gynaecological surgery and miscarriage after miscarriage. Monroe regularly spoke of her longing for children, and made a point of favouring children's charities and funds for orphanages. At her funeral, instead of flowers, donations would be directed to children's hospitals, and a bequest continues to go to a London children's clinic. The accounts given of Monroe's physical disorders represent perhaps a more general problem with men's inter-pretations of women. No biography of Marilyn Monroe contains an adequate appreciation of the effect these illnesses had upon her outlook, her behaviour (particularly her 'lateness'), or her own aspirations.

Sometimes in her starlet days period pain would cause Monroe to stop her car abruptly, jump out, and crouch on the ground in agony. Zolotow (1961) recalls visiting her studio dressing room and noting at least fourteen boxes of pills – nearly all were painkillers for menstrual pains. Monroe's Los Angeles physician, Dr Lee Siegel, has identified her ailment as endometriosis, a condition in which endometrial tissue (the lining of the uterus) spreads through the abdomen. Extreme pain during menstru-ation, and pain in the reproductive organs, is a usual symptom. Amy Greene sometimes found Monroe screaming in agony from which pills brought no respite. Women who have endometriosis and want to bear children are urged not to wait too long before becoming pregnant, for the disease is progressive and worsens with time (see Berkow 1977).

Henry Rosenfeld, a dress manufacturer who was a close friend of Monroe, told Summers (1985: 23): 'She wanted a baby so much that she'd convince herself of it every two or three months. She'd gain, maybe fourteen or fifteen pounds. She was forever having false pregnancies.' When Monroe was in hospital during 1952 to have her appendix removed, a nurse found this note, written in pencil, taped to the unconscious patient's stomach:

Most important to Read Before *operation*

Dear Doctor,

Cut as little as possible I know it seems vain but that doesn't really enter into it – the fact that I'm a *woman* is important and means much to me. You have children and you must know what it means – *please Doctor* – I know somehow you will! thank you – thank you – for Gods sakes Dear Doctor No *ovaries* removed – please again do whatever you can to prevent large *scars*. Thanking you with my all *heart*.

Marilyn Monroe.

During Monroe's final few movies, her unpunctuality and unpredictable moods were significantly affected by these physical agonies and anxieties. Some men seem to regard Monroe's menstrual pains as nothing more distracting than a mere headache. For biographers to mention such factors, as they do, and then to effectively judge them as not an excuse for 'Marilyn's undependability', is a poor reflection on these writers' interpretive abilities. Before beginning shooting *Some Like It Hot* in 1958, Monroe asked her friend Norman Rosten (1980: 71):

> Should I do my next picture or stay home and try and have a baby again? That's what I want most of all, the baby, I guess, but maybe God is trying to tell me something, I mean with my pregnancy. I'd probably make a kooky mother; I'd love my child to death. I want it, yet I'm scared. Arthur [Miller] says he wants it, but he's losing his enthusiasm. He thinks I should do the picture. After all, I'm a movie star, right?

The movie star made her movie and lost her baby. Her marriage to Arthur Miller began to distintegrate. By 1960, Monroe would often stand in a telephone booth near a park and watch children playing. According to several of her biographers (Summers, Guiles, and others), Monroe became pregnant during the last year of her life, whilst she was deeply involved with Robert Kennedy. In June 1962 she told photo-journalist George Barris: 'A woman must have to love a man with all her heart to have his child. I mean, especially when she's not married to him. And when a man leaves a woman when she tells him she's going to have his baby, when he doesn't marry her, that must hurt a woman very much, deep down inside.' (in Wilson 1962) The pregnancy was terminated in July – it had been tubular, as had her previous pregnancies. At the same time as her conversation with Barris, she told W. J. Weatherby (1976: 202–3) of her current lover: 'Only problem is, he's married right now. And he's famous, so we have to meet in secret. . . He's in politics. . . In Washington.'

When, at the peak of her fame, Monroe was advised to enter a psychiatric clinic to recuperate from her 'nervous' illness, she was shocked at her treatment. She heard iron doors slam behind her, she saw bars on her windows, a doorless toilet, and a glass pane on her door. 'What are you doing to me?' she screamed. Such experiences would surely disturb the stablest of personalities; with hindsight, it seems impressive that someone with a fear of mental illness kept control for so long. There was not anything 'inevitable' about Monroe's decline, although the biographies seize on incidents that can be slotted into the 'Norma Jeane's inheritance of insanity' thesis. What is indubitably extraordinary is the press coverage of such incidents. The *New York Journal – American* (6 March 1961) reported Monroe's release from the clinic with typical tastelessness: 'All's well with the world, men, so fear not, fear not. Marilyn's face still has the ethereal rose-petal texture, the smile's as delicately soft as ever, the figure – ah yes, the figure – and best of all they've untied the knots in her nerves.'

Throughout Monroe's illnesses, she was obliged to endure a rigorous filming schedule. In spite of her huge success, Monroe had been tied to a Fox contract that paid a maximum of $1,000 a week. Despite the title of her film *Gentlemen Prefer Blondes,* Monroe received $18,000, compared with co-star Jane Russell's $100,000. Monroe had to pay taxes, agent fees, her acting coach Natasha Lytess, and beauticians. In a bid for independence, Monroe agreed to form a production company, Marilyn Monroe Productions, with photographer Milton H. Greene. She flew to New York in December 1954, travelling under the pseudonym of Zelda Zonk. By mid-December 1955, negotiations with Fox were nearing settlement: Monroe had director approval, an obligation to make only four films in seven years (with the right to make one film per year for an outside studio), the right to veto 'substandard' screen-plays, and a salary per film of $100,000 with a percentage of the profits. Monroe signed her new contract at the end of 1955, and could look forward to a potential income of some $8 million over the next seven years.

Monroe enjoyed the New York environment. Now married to Arthur Miller ('Egghead Marries Hourglass' sneered the press), she became involved in political discussions. The House Un-American Activities Committee (HUAC) had summoned Miller, author of the anti-witchhunt play *The Crucible,* in an attempt to recapture the front page. Miller was offered 'an easy ride' if he would arrange for his then-fiancée, Monroe, to pose for photographs beside Congressman Walter. Miller refused and was convicted; it took two years to gain an acquittal. Throughout his campaign he enjoyed the public support and financial assistance of Monroe. She told W. J. Weatherby (1976: 54): 'Some of those bastards in Hollywood wanted me to drop Arthur, said it would ruin my career. They're born cowards and want you to be like them. One reason I want to see Kennedy win is that Nixon's associated with that whole scene.' It was ironic that Monroe was, and is, cast as the cause of Miller's 'creative silence' at this time.

Through Miller, Monroe met Cheryl Crawford, a producer and director for the stage. Crawford was impressed with Monroe's desire to play 'serious' roles, and recommended she contact the Artistic Director of the Actors Studio, Lee Strasberg. Strasberg was the most notorious figure in American theatre at the time, popularizing a 'Method' of acting based on Stanislavsky. Method actors were encouraged to summon up personal inner feelings and memories that could facilitate the 'performance' of a role. Marlon Brando, Paul Newman, and Montgomery Clift were famous members of the Actors Studio, and Monroe was given special access whenever she was available. Strasberg probed to discover why Monroe could not express herself clearly. People kept misinterpreting what she said, Strasberg was told, and they laughed at her or observed her with strained expressions. Monroe said that she had reached the point where she could not really say what she meant, because she had to think carefully over every word lest she be ridiculed. As Strasberg grew to know Monroe, he said he had seldom seen anyone who possessed a better, more direct control of language. She sat in on classes, and once performed a scene from *Anna Christie;* the consensus was that her work was sensitive, hushed, and tremulous. Strasberg told his class that Monroe could be a fine Grushenka in *The Brothers Karamazov* or a Cordelia in *King Lear.* Miller's screenplay and Strasberg's students combined in 1961 to make a 'Method film', *The Misfits.* Her self-confidence was shaken at this time by her association with Frank Sinatra. According to his biographer Kitty Kelley (1986), Sinatra would shout at Monroe in public: ' "Shut up, Norma Jean. You're so stupid you don't know what you're talking about".' However, other people found Monroe an intelligent and attentive companion. Clifton Webb wrote in *Picturegoer* (11 June 1955) that Monroe was 'very sweet, very serious. She likes to talk about the theatre and the kind of thing that makes people tick. She is intense and completely straightforward. She reads all the time. She is in complete earnest towards her career.' Before her last illness, plans were at an advanced stage for Monroe to play the part of Sadie Thompson in a television dramatization of Somerset Maugham's *Rain.* Eleanor Powell argued that Monroe 'wasn't at all like the "sex-pot" role she portrayed on the screen. She was an extremely intelligent woman and a fine actress.' John Huston's plan to make *Freud,* from a screenplay by Jean-Paul Sartre, was also being discussed (*Los Angeles Mirror,* 20 April 1961; Huston, 1981). Sartre, a great cinephile, thought Monroe 'one of the greatest actresses alive', and she was first choice to play the part of Cecily, one of Freud's psychotic patients. Saul Bellow, accompanying Monroe to the Chicago premiere of *Some Like It Hot,* described her in correspondence with the author (4 November 1986) as 'a very witty woman'.

Monroe's New York friends found her to be very politically aware. Joan Greenson (in an unpublished manuscript) recalls: 'She identified strongly with the workers, and she always felt they were her people.' Arthur Miller, in his *After the Fall* autopsy on his marriage to Monroe, asked: if she was for the workers, why did she fire so many? Coming from a man whose work embraces the necessarily contradictory in people, this response seems conveniently out of character. In her interview with *Life* journalist Richard Meryman (3 and 17 August 1962), Monroe said: 'If I am a star – the people made me a star, no studio, no person, but the people did.' W. J. Weatherby (1976: 129–31) remarked to her how so many black American women seemed to admire her; Monroe replied.

> It's easy to understand the slave system when you've been through the star system. . . I don't want to be a symbol of anything. Negroes can sometimes see through appearances better than whites. Blondes don't even appeal to some of them. I'm not a sex symbol, but busty Miss Anne.

She went on to say:

> I don't know about politics. I'm just past the goodies and baddies stage. The politicians get away with murder because most Americans don't know any more about it than I do. Less even. Arthur was always very good at explaining, but I felt at my age I should have known. It's my country and I should know what they're doing with it.

In 1960, Monroe sponsored the National Committee for a Sane Nuclear Policy. When she expressed her loathing for the atom bomb to Robert Kennedy, he accused her of 'turning communist'. A registered Democrat, she was named as an alternative delegate for Roxbury, Connecticut, in the primaries. An American leftist, Fred Vanderbilt Field, came to know Monroe and would invite her to meetings and social occasions. He recalls: 'She told us of her strong feelings for civil rights, for black equality . . . her admiration for what was being done in China, her anger at red-baiting and McCarthyism, and her hatred of J. Edgar Hoover.' When Ella Fitzgerald was barred from performing at a Los Angeles night club, Monroe promised to sit at the front table each night if the singer was employed. Fitzgerald says, 'Marilyn was there, front table, every night. . . After that, I never had to play a small jazz club again.' The unusually public identification Monroe made with oppressed groups encouraged the FBI to start a '105' file on her from 1955; such a file technically applies to 'foreign counter-intelligence matters'. J. Edgar Hoover probably used the 105 file as a pretext to investigate any politically concerned person. Monroe's file contains thirty-one pages: only thirteen have been released, heavily censored.

The marriage of Monroe and Arthur Miller finally collapsed during the making of the *The Misfits* in 1961. In 1962, reluctantly beginning the filming of *Something's Got to Give*, Monroe was in considerable physical and emotional pain. Always a shrewd judge of scripts, Monroe was thoroughly unimpressed by the current work, and she was further dismayed when rewritten scenes were delivered to her each morning. She became determined to extricate herself from the production. Movie executives pictured her as 'severely mentally disturbed'; this is untrue, she was in fact reacting in fairly typical fashion to what she considered an outrageous situation. She told Richard Meryman that before his interview would be published, 'the legend may become extinct. . . Not the girl, but the legend.' The studio fired Monroe on 9 June 1962 for failure to fulfil her contractual obligations. Fox, reading Monroe's behaviour as a repetition of the problems during *The Misfits,* had failed to realize it had compounded these problems by further irritating her in its total disregard for her as a serious performer. The studio (financially weakened by Elizabeth Taylor's epic *Cleopatra* and by the competition from television) had treated Monroe as a mere commodity, and Monroe rebelled in the only ways open to her.

Two months after being fired, Monroe was dead. On the night of 4 August 1962, the 36-year-old star was found lying nude in the bedroom of her half-furnished home, one hand tightly-clasping the telephone, killed by an overdose of barbiturates. Until Summers' 1985 account, it was said that at 3.30 a.m. on 5 August 1962, Mrs Eunice Murray, the housekeeper, awoke, noticed the telephone cord leading under Monroe's bedroom door, and became concerned. Monroe's psychiatrist, Dr Ralph Greenson, was called, and he found his patient dead. After an examination by her personal physician, the police were summoned at 4.25 a.m. By 9.00 a.m. an ambulance delivered the body to the Coroner's office. After the autopsy at 10.30 a.m., it lay unclaimed on a slab in the storage vault, as Coroner's Case No. 81128. Monroe dead had no one to claim her, her life ending as it had begun.

In the new version by Summers (based on newly released evidence), it is alleged that Monroe was having an affair with both Kennedy brothers; that the death was probably covered up for several hours in order to allow Robert Kennedy to leave Los Angeles; that teamster boss Jimmy Hoffa was out to get Robert Kennedy and used the Monroe affair to do so; and that Monroe may, possibly, have been murdered. In 1982, the Los Angeles district attorney's office reinvestigated the case, concluding: 'Her murder would have required a massive in-place cover-up' including witnesses, 'the actual killer or killers', the coroner's office, and virtually the entire Los Angeles police department. The allegations were dismissed, but Summers showed there were still grounds for doubt.

Monroe certainly made repeated and unsuccessful attempts to tele-phone Robert Kennedy at the justice department in July 1962. Her old friend Robert Slatzer has claimed that she recorded Kennedy's 'pillow talk' in a diary, including details of the CIA plot to assassinate Castro using Mafia leaders. Slatzer (1975) says that he warned Monroe the diary was a 'time bomb', and she told him of her plan to hold a press conference, 'blowing the affair wide open' unless Kennedy contacted her. Slatzer advised her to try to forget what she herself had called 'a bad experience', but she was determined to confront Kennedy. Two days before the press conference, says Slatzer, Monroe died. According to Summers, on her last evening alive, Monroe phoned the Kennedy confidante, Peter Lawford, telling him: 'Say goodbye to the President.' Then, according to witnesses, news of her death was withheld from police for five hours while any evidence connecting her with Kennedy was removed from her home. Eunice Murray, after twenty-three years, has changed her story and now insists that she found the body at midnight. She says 'Why, at my age, do I still have to cover this thing?' (Summers 1986: 515). The owner of the ambulance service told Summers that a comatose Monroe was taken shortly after midnight to hospital, where she died, and her body was then illegally returned to her house. Kennedy, it is claimed, flew away by helicopter.

Although the plumbing was switched off and no liquids were found in Monroe's room, she is supposed to have swallowed vast amounts of tablets. Two kinds of drugs were found in her system during the autopsy – nembutal in a lethal amount and chloral hydrate in a toxic amount. Dr Keith Simpson is quoted by Summers as saying that the fatal dose could have been inserted colonically – i.e. by enema – which would have left no marks. Enemas were a common fad amongst Hollywood people as shortcuts to weight loss. Peter Lawford's former wife, Deborah Gould, claimed to Summers that her husband said 'Marilyn took her last big enema.' Lawford is not alive to explain this comment. Regardless of how or why the drugs were taken, it is evident that Monroe lived for several hours after ingesting them – long enough to have been saved if the signals she gave to those she phoned had been picked up and acted upon.

Summer's sources include Slatzer's private detective, Milo Speriglio. In Speriglio's own account, *The Marilyn Conspiracy* (1986), the murder thesis is the most convincing available. Certainly Speriglio's investigations caused considerable controversy amongst the Mafia. Robert Slatzer's publisher was badly beaten up, and Slatzer has required police protection over a number of prolonged periods: he has a recording of a telephone conversation with an anonymous caller, informing him of a contract out on his head. An ABC News programme, *20/20*, produced a special feature on Speriglio's and Summers' findings: station chief Roone Arledge, a friend of Robert Kennedy's widow, Ethel, decided to ban the

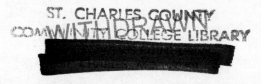

programme at the last possible moment. Speriglio's argument, compromised as it is through certain unidentified sources, serves as testimony
to the seriousness of this matter.

The cast of characters in this extraordinary drama seem tailormade for
such intrigue. Dr Ralph Greenson was Monroe's psychiatrist, Dr Hyman
Engelberg was her doctor; whilst the former urged the actress to avoid any
barbiturates, the latter encouraged her to continue taking them – Monroe
trusted them both. When Sgt Jack Clemmons of the West Los Angeles
Police arrived after being informed of Monroe's death, he reported
finding a 'smirking' Greenson and a 'remorseful-looking' Engelberg by
her bedside. Clemmons, aware of the usual disorder when the victim
undergoes a convulsive overdose, was surprised at the scrupulously tidy
state of the room and the strangely 'swan-like' position of the dead body.
Monroe's housekeeper, Mrs Eunice Murray, struck Clemmons as being
curiously 'vague' and 'nervous'. The scene seems ever more suspicious
when one learns that Dr Greenson treated both Monroe and her 'Mafia'
link, Frank Sinatra, and that these two stars shared the same lawyer,
'Mickey' Rudin – who was Greenson's brother-in-law. Greenson had also
introduced Mrs Murray into the house, as a long-time friend of the
psychiatrist. In the 1980s, Mrs Murray changed her story (as noted
earlier) and now insists that Monroe had been found dead over three
hours before the police were notified. Patricia Newcomb, Monroe's
personal assistant, refuses to recall her memories of that night in August
1962, other than to stress that Robert Kennedy was nowhere near the area
at the important time. Mrs Murray *now* is adamant that Kennedy had
been there, arguing with Monroe. Newcomb, after the demise of her
employer, entered into a series of working relationships within the
Kennedy entourage.

Speriglio contends that Monroe's death should be classified as
homicide rather than the existing verdict of 'probable suicide'. Sam
Giancana ordered the hit, says Speriglio, Johnny Roselli planned it.
Although a massive overdose of barbiturates was taken, Speriglio repeats
that there were no signs of water vessels in the bedroom, and the
plumbing had been switched off for the renovation of the bathroom. The
fatal dosage, suggests the author, (agreeing with Summers), could have
been administered through the application of 'protoclysis', an enema. If
the poison had been administered in this way, it would not have reached
the stomach. It would only have gone into the intestine – and that, by
some quirk of the Coroner's Office, was never tested. By murdering
Monroe, it is argued, Robert Kennedy's enemies would be in a position to
achieve their twin objective: Monroe's death by 'overdose' connected to
her romance with Kennedy would at once politically cripple the Attorney

General and eliminate what the mobsters regarded as nothing more than an overtalkative, unreliable 'dumb blonde'. According to Speriglio, the 'Monroe–Kennedy' tapes included a quarrel in which Monroe told Robert that she was tired of being 'passed around like a piece of meat', and that she had had it, and did not want Jack or Bobby to 'use' her anymore. Mafia bosses Sam Giancana and Jimmy Hoffa mixed with Monroe and Sinatra at the Cal-Neva Lodge gambling location. If her plan to hold a press conference to divulge her knowledge of the Kennedy's political dealings was indeed a genuine one, it must have been evident to all concerned what a 'time-bomb' this was: John Kennedy, not Richard Nixon, might have become the first US president to resign from office.

The glib explanations of Monroe's slide into a suicidal state are losing their supporters. Sam Cordova, foreman of a grand jury asked to review Monroe's death, in turn asked for a special investigator to enquire into her case. 'There is enough evidence to substantiate a special prosecutor to work with the grand jury on an investigation.' Cordova said in 1985, 'A full investigation has never been done by the grand jury.' The ubiquitous Slatzer points out that the cause of death on the death certificate has changed from 'suicide' to 'possible suicide' and then to 'probable suicide'. 'People have not testified under oath,' adds Cordova. 'This should have been done a long time ago.' A few months after her daughter's death, Monroe's mother escaped from her sanitarium and was found by a clergyman in a Baptist church. 'Marilyn', she said (not even calling her by her 'real' name) 'she has left. . . People ought to know that I never did want her to become an actress. . . Her career never did her any good' (in Preston 1963).

There was world-wide sorrow – and then an outcry: 'Hollywood' was blamed. *Pravda* made the observation that Monroe was the victim of a decadent society. The Berlin newspaper *Der Vend* declared that Hollywood created and destroyed her. It said her life was 'a tragedy of the dream factory'. The Vatican's *L'Osservatore Romano* made a typically condescending proclamation: 'We hope that, in the desperate solitude of this poor woman, and at the last moment, there was Someone who during life was kept away, and (we wish) that hope and peace has made the dying actress smile.' When Arthur Miller was asked if he would attend Monroe's funeral, he replied, 'Why? Will she be there?'

W. J. Weatherby (1976: 229), revisiting the places where he once spent time with Monroe, notices the altered buildings, fashions and street sounds, but then adds: 'But at least one person would never change now, would never grow old'. Guiles makes his conclusion patently clear long before *his* 'Marilyn' dies; Part two is titled 'Goodbye, Norma Jeane'. After closing with criticisms of the studios and the demanding public, the (unnamed) 'important Senator', her unresponsive friends, Guiles clearly avoids upsetting his 'tragic' composition: 'Yet her early death was inevitable, no matter what the immediate circumstances surrounding the act were... Marilyn was destroyed by a great many events, beginning with the circumstances of her birth.' Thus, Guiles seems to be sponsoring one of the 'purest' examples of tragedy in human history. His book is an essential source, and thus his thesis seeped deeply into subsequent interpretations. Two 'biopics', *Goodbye Norma Jeane* (1976) and *Marilyn: The Untold Story* (1980), merely feed off the conventional myths, and offer a predictable thesis (coming from the movie industry) that 'Marilyn' was doomed long before Hollywood began 'irritating' her.

The evidence of the implicit (and sometimes blatant) sexism in these post-mortems is considerable. Hoyt (1967: 11) makes this extraordinary *non sequitur:*

> The study of Marilyn's life convinced me that, in the essentials of the American dream, women are more idealistic than men, they honour accomplishment for its own sake more than men. Or perhaps they are more naive than men and not so convinced of the basic evil of mankind. Marilyn gave evidence of conviction that mankind was evil, and yet she never ceased to place her trust in her fellow man.

Some Two hundred pages later, Hoyt concludes (p. 249):

> Marilyn was innocent, innocent of intent to do evil and innocent of understanding of the world's ways, and this was her tragedy.

This tragic romanticism was still operative in the 1970s pop songs dedicated to Monroe: Elton John and Bernie Taupin produced 'Candle in the Wind' (1974), overtly influenced by Guiles, which neatly captured the version of Monroe as the submissive woman: 'They set you on the treadmill/And they made you change your name,' 'And it seems to me you lived your life/Like a candle in the wind.'

Anthony Summers (1985: 367–8) concludes with a remark situating Monroe alongside other movie stars and noting her place in the 'Age of Anxiety'. He suggests that 'Everyman had made her a goddess.' Norman Mailer (1973: 242), who did not research the 'Kennedy connection' methodically enough to protect himself from large-scale criticisms, moves toward a conclusion which is self-consciously provocative and playful, thinking out loud: 'Why then not also see her in these endlessly facile connections of the occult as giving a witch's turn to the wheel at Chappaquiddick?'

Writing is always rewriting. Anything 'enigmatic' in a life is, for the biographer, a challenge to be overcome and 'explained away'. To read a history is thus also to read another reading. As Barthes (1975: 64) came to appreciate:

> Text means Tissue: but whereas hitherto we have always taken this tissue as a product, a ready-made veil, behind which lies, more or less hidden, meaning (truth), we are now emphasizing, in this tissue, the generative idea that the tissue is made, is worked out in a perpetual interweaving; lost in this tissue – this texture – the subject unmakes himself, like a spider dissolving in the constructive secretions of its web.

It should have been evident during this survey of the biographical 'mediation' of Marilyn Monroe how polysemic is each 'finished' text, how manifold are the meanings as the writer works towards closure whilst the material breaks out of every boundary. Julian Barnes has his character in *Flaubert's Parrot* (1984: 38) express this dialectic:

> The biography stands, fat and worthy-burgherish on the shelf, boastful and sedate: a shilling life will give you all the facts, a ten pound one all the hypotheses as well. But think of everything that got away, that fled with the last deathbed exhalation of the biographee. What chance would the craftiest biographer stand against the subject who saw him coming and decided to amuse himself?

W. J. Weatherby suffered this uneasiness when, talking with Monroe, he found her noticing his notebook; she felt suspicious, he felt defensive. She suddenly smiled, squeezed his hand, and said (1976: 128–9): 'You're the first shy reporter I've ever met. You're like me – an old softie. . . Put me in the notebook if you like, but don't write about it now. Do it when I retire!'

The books on Marilyn Monroe do not contain 'the real Marilyn', signed, sealed, delivered and yours. The human being that was Marilyn Monroe cannot be reconstructed by piecing together the black and white marks in the text, but what can be reanimated is the cultural praxis at work in each interpretation, in each word-picture. Certainly, one might point to a multiplicity of 'Marilyns': however, to hold that a given text is 'different for everybody' is as sociologically blind as to hold that it is the 'same for everybody'. The latter implies the possibility of a 'definitive' analysis (*here* she is!), capable of determining the use-value of the text in abstraction from the context of its use; the former position implies a malleable transparency of the text to the determinations of the particular audience – removing any practical basis for supporting through analysis any text against any others.

'Please don't make me a joke,' said Monroe. What *do* we make of her, and how? The proposition that texts can be manoeuvred into practically any ideological space is itself quite abstract in the idea of freedom it posits, its occlusion of the determinacy at any moment of specific institutions of meaning, within the spaces of which are given the grounds of contradiction, mobilization, and reappropriation. Who is writing about Monroe? What for? Who for? Why do they select *that* fact and discard another? Monroe is a significant subject for sociological study, raising questions of gender and of celebrity, of the role of popular icons in the marketplace and in the memory. Yet she is also significant in herself, for the personality evident in the images and the interviews. It is important to retain a sense of the tragedy involved when anyone dies so young and so alone in such dubious circumstances. There seems a frightening continuity between many 'serious' studies and the one-dimensional biographies they are meant to supercede. A more sensitive approach is needed to overcome some fundamental contradictions: that between outside and inside, public and private, the sociological and the psychological, between my being-for-others and my being-for-myself, objectivity and subjectivity – between society and the monad. As we finish, Sartre's Roquentin struggles on, never quite seeing the plot he is a part of:

> I am beginning to believe that nothing can ever be proved. These are
> reasonable hypotheses which take the facts into account: but I am only
> too well aware that they come from me, that they are simply a way of
> unifying my own knowledge. . . Slow, lazy, sulky, the facts adapt
> themselves at a pinch to the order I wish to give them, but it remains
> outside of them. I have the impression of doing a work of pure
> imagination. And even so, I am certain that characters in a novel
> would appear more realistic, or in any case would be more amusing.
> (Sartre 1976: 26)

Monroe by Beaton: 'the wonder of the age'.
Photograph by Cecil Beaton courtesy of Sotheby's, London.

3

Marilyn in Focus

Photography has the same relation to History that the biographeme has to biography.

Roland Barthes

People had a habit of looking at me as if I were some kind of mirror instead of a person. They didn't see me, they saw their own lewd thoughts, then they white-masked themselves by calling me the lewd one.

Marilyn Monroe

Marilyn Monroe was one of the most photographed individuals of the twentieth century: a photograph of her broke the UK record when it was sold in 1986 for £17,600 at a Sotheby's auction. (The *Guardian* 26 April 1986 and the *Observer* 23 March 1986) Each year it seems yet another 'unknown' image of Monroe is unearthed and exhibited. One more, just another, another angle, another expression, and the collector will *have* her, possess the whole person. Yet, another is never enough, and the magic that was 'Marilyn' remains elusive, evident but enigmatic. In magic one needs only an object belonging to the person in order to bewitch them: Pythagoreans used to obey the injunction to smooth the bed soon after rising to remove the imprint of the body so that it could not be used to the owner's detriment; in popular romance, a kiss would be treasured and the area kissed left unwashed; in motion pictures, the most arresting images are frozen into stills. In photography, in focus, can we see the real Monroe? In the early 1970s, Larry Schiller launched an exhibition in the States entitled 'Marilyn Monroe – The Legend and the Truth'; it comprised over 16,000 photographs by several leading photographers, and formed the basis of Norman Mailer's *Marilyn* book. Although the exhibition certainly displayed the 'legend' of Monroe, it was hard to locate the 'truth'.

The British photographer Cecil Beaton met Marilyn Monroe in the mid-fifties in order to produce a picture spread for *Vogue* magazine. 'This then,' he wrote (Beaton 1957), 'is the wonder of the age – a dreaming somnambule, a composite of Alice in Wonderland, Trilby, and a Minsky artist. Perhaps she was born the postwar day we had need of her. . . Like Girandoux's *Ondine,* she is only fifteen years old . . . as spectacular as the silvery shower of a Vesuvius fountain, an incredible display of inspired, narcissistic moods.' Beaton was not untypical in his fascination for Monroe: the leading photographers of her time helped to elevate her to that elite body of the media legends. Monroe was celebrated as a 'sex symbol', yet so many of her admirers could only turn to tautology: Arthur Miller spluttered, 'She's all woman, the most womanly woman in the world.'

Select a typical photograph of Marilyn Monroe. Study it, scrutinize its features, sense its execution. Was this one pose unique, snapped between its forming and dissolving, or was it an achingly familiar pose, shot successively, here as merely one amongst many? Monroe, as a still image, excites the most intense reaction from the spectator: what happened before, after, who was in the room with her, who was behind the camera, what was she thinking, saying, feeling? The single image sparks off a series of questions (every caption trails off into three dots. . .). A typical pose displays her with glistening blonde hair, brightly dressed, her face set with half-closed eyes and half-open mouth (not quite seeing, not quite speaking) – a picture of sheer promise. Parts of her body – the dark red lips, the long black eyelashes, the shaped and elongated eyebrows – are heavily outlined, whilst other parts – her neck and torso – are traced with trinkets and clothing that accentuate the roundness and fullness of her figure. The bent-forward attitude of her head and upper body, the slanting seductiveness of the eyes and lips, fashion an alluring self-image quite distinct from the cryptic off-camera self Monroe spent her life striving to define. As seen in portraits, Monroe intrigues us, invites us to ponder on our peculiar relationship to her.

What is the relationship between spectatorship and representation? What does looking have to do with sexuality? Why are women's bodies so prominent in consumer society? In this chapter these questions will be addressed through the consideration of the woman who was called 'Hollywood's number one sex symbol' and 'the most looked-at woman in the world'. The photographic image of Marilyn Monroe inhabits various contexts: cultural contexts of spectatorship; institutional, social, and historical contexts of production and consumption. When she started to pose, she was shown as 'cheesecake', a pin-up. The pin-up's singular preoccupation with the female body is tied in with the project of defining the 'true' nature of female sexuality. Femaleness and femininity are constructed as a set of bodily attributes reducible to a sexuality that puts itself on display for a masculine spectator, allowing him space for his fantasies. In these ways the pin-up invites the spectator to participate in a masculine definition of feminity.

Marilyn Monroe was presented as '*The* American Woman'. The bureaucratic regulation of populations takes place through the individuation of bodies and in contemporary societies the moral regulation of bodies is brought about under the auspices of health. The cultural significance of the female body is not only (not even primarily) that of a flesh-and-blood entity, but that of a *symbolic construct*. Everything we know about the body – as regards the past, and even as regards the present – exists for us in some form of discourse; discourse, whether verbal or visual, fictive or historical or speculative, is never unmediated, never naive. The more technology develops the diffusion of information (and notably of images), the more it provides the means of masking the constructed meaning under the appearance of the given meaning (see Colmar 1979). It would be wrong, however, to entertain the dream of unpeeling the several social skins that cover the object of inquiry; it is quite illegitimate to attempt to give the myth of Marilyn Monroe 'a good dressing down'.

Dress links the biological body to the social being, and public to private – thus forcing us to recognize that the human body is more than a biological entity. It is an organism in culture, and its boundaries are unclear:

> Can we really assume that the limits and boundaries of the human
> body itself are obvious? Does "the body" end with the skin or should
> we include hair, nails. . . What of bodily waste materials?. . . Surely the
> decorative body arts such as tattooing, scarification, cranial
> modification and body painting should also be considered. . . [and] it
> has been shown that it is insignificant (if not inaccurate) to sharply
> differentiate between bodily decoration and adornment on the one
> hand and the clothing of the body on the other hand. (Polhemus
> 1978: 28)

In all societies the body is 'dressed', and everywhere dress and adornment play symbolic, communicative and aesthetic roles. Dress is 'unspeakably meaningful'. 'Fashion' is dress in which the key feature is the rapid and regular changing of styles; in modern Western societies fashion sets the terms of all sartorial behaviour. The mass production of fashionable styles – itself highly contradictory – links the politics of fashion to fashion as art. It is connected both to the evolution of styles that circulate in 'high' and avant-garde art; and to popular culture and taste. As with any other aesthetic enterprise, fashion may be understood as ideological, its function being to resolve formally, at the imaginary level, social contradictions that cannot be resolved (see Jameson 1981: 79). Fashion is thus essential to the world of modernity, the society of spectacle and surveillance.

Marilyn Monroe entered films via fashion. Hollywood and fashion always enjoyed a close relationship, with several bosses having their roots in the clothing trade. Monroe, struggling to succeed in the 'Dream Factory', found it easiest to begin her career as a model, using her body as an image for consumption. The movie sex symbol had long been imagined when Monroe came into sight: Jean Harlow was a rather disturbing 'Red Woman', Betty Grable was a wartime 'White Woman'. Monroe was drawn into a preformed pattern of sexual iconography; that she refused to submit to its conventions has been part of her distinctiveness. The process whereby Marilyn Monroe acquired her public image was a long and complex affair, a well-practised ritual in which she and her specialists indulged in a collaborative cultural conceit. Allan 'Whitey' Snyder, Monroe's regular make-up man, told Zolotow (1961: 102):

> Marilyn has make-up tricks that nobody else has and nobody knows. Some of them she won't even tell me. She has discovered them herself. She has certain ways of lining and shadowing her eyes that no other actress can do. She puts on a special kind of lipstick; it's a secret blend of three different shades. I get the moist look on her lips for when she's going to do a sexy scene by first putting on the lipstick and then putting a gloss over the lipstick. The gloss is a secret formula of vaseline and wax. You see, I'll say "Kiss me, honey", and when she puckers her lips I put on the gloss. Interesting thing about Marilyn is she's one of the few gals you can photograph full face and she'll look good, the most of them you take it three quarters or side view. Her left profile is great in three quarters. Her right profile is bad, for some reason. She's got a bad jawline on that side. If she has to work a scene with her right profile to the screen, we have to do a lot of work on the right jawline.

Monroe's platinum blonde hair was made with Snyder's own blend of sparkling silver bleach plus twenty-volume peroxide and a special formula of silver platinum to take the yellow out. During filming this solution would be administered once every four days. Hairdresser Gladys Whitten would supervise Monroe's coiffure, and also paint her fingernails and toenails with platinum polish. Snyder powdered Monroe's shoulders, pencilled lines around her eyes, and added high gloss on her lips. For particularly prominent showbusiness occasions, Monroe's evening dress would be so designed as to need her to be sewn into it. The whole process could often take over six hours.

Monroe's supposed 'narcissism' was significantly a narcissism *pour autres:* her photographic image was dressed up with somewhere to go – your gaze, your voyeuristic pleasure. She was a *bright* star ('first star you've seen tonight'). Christopher Lasch (1979: 167) notes that one prevalent response to the 'spiritual desolation' of modern life is a kind of consumerist narcissism:

> To the performing self, the only reality is the identity he can construct out of materials furnished by advertising and mass culture, themes of popular film and fiction, and fragments torn from a vast range of cultural tradition. . . In order to perfect the past he has devised, the new Narcissus gazes at his own reflection, not so much in admiration as in unremitting search of flaws. . . Life becomes a work of art. . . All of us, actors and spectators alike, live surrounded by mirrors.

The 'lipstick, powder and paint' of consumer society cannot simply be wiped away from the social body. When Mary Quant marketed a foundation cream called 'Starkers' advertisements displayed an apparently naked young woman, crouched clasping her knees so that she was decently cloaked in curtains of Lady Godiva hair – but it was still make-up, not nature.

The notion of social construction is based on the view that at birth a baby has the potential to develop in a variety of ways, limited to some extent by genetic heritage, but equally dependent upon the environmental influences that shape its experiences and provide a comparatively favourable or unfavourable situation for growth. Many of the key aspects of this development occur in early childhood. By the time we become adults, therefore, our capacity to choose freely is greatly restricted by the way in which our personality has developed. It is significantly influenced by external circumstances such as class, gender, age, and locale. The pseudo-democracy of mass communications in the inter-war period seemed to offer all women the 'right' to femininity. In theory, the fashion magazines, diets and cosmetics promised every woman film star looks. Monroe recalled: 'I dreamed of myself becoming so beautiful that people would turn and look at me when I passed.' The cinema made cosmetics not only desirable but respectable. Yet when every woman could paint a mask of fashionable beauty on to her face, the democracy of beauty failed to appear.

'Cosmetic' became, for critics, synonymous with superfluity. As Monroe began her career, the critics were focusing on the shallowness of culture in a time of international unrest. In the late 1940s, Godfrey Hodgson has written, there was 'an almost operatic sense of coincidence, a crescendo of foreboding' – with the Berlin blockade and the death of Jan Masaryck (1948), the fall of Shanghai (May 1949), the establishment of Mao as head of a Chinese People's Republic (September 1949) and, in 1950, the conviction of Alger Hiss for perjury. Marilyn Monroe said to reporters, 'My nightmare is the H-bomb. What's yours?' (in Redbook August 1962). The bomb dropped on Bikini was covered with a picture of Rita Hayworth. Soon after the announcement, in 1949, that the Russians had the bomb, the editors of the *Bulletin of the Atomic Scientists* moved the hands of the clock that regularly appeared on its cover to three minutes to midnight. It was a startling image of the imminence of nuclear war. The nuclear bomb was exploded over Hiroshima at 8.15; all subsequent advertisements for clocks and watches in America depicted hands at 8.20. 'Fashion', as Leopardi reminds us, 'is the mother of death.'

Within the USA, however, there was general confidence and optimism, a belief that abundance and 'revolutionary capitalism' could eliminate social barriers, creating 'a nation of the middle class'. In *Capitalism, Socialism and Democracy* (1942), Joseph Schumpeter predicted that the creation of new resources might 'annihilate the whole case for socialism'. J. K. Galbraith, in *American Capitalism* (1952), declared: 'It works, and in the years since World War II, quite brilliantly.' In the book which gave a catchphrase to the whole phenomenon, *The Affluent Society* (1958), Galbraith claimed: 'Production has eliminated the more acute tensions associated with inequality.'

In the 1950s, the American manufacturers perfected the technique of 'designing the consumer to the product', instilling in people a 'desire for possession' and creating an 'image' into which people felt eager to fit. Arthur Drexler, the automobile designer, had suggested that the designer needed 'the training and inclinations of a psychoanalyst'. *Look* magazine, introducing 'the Fabulous Fifties look', the 'age of everyday elegance', had suggested that 'Functionalism today is not enough for Americans.' Designers must now offer 'Plush at popular prices'. If manufacturers were going to try to influence the consumers, as much needed to be found out about them as was possible. Market research was the American pseudo-science that purported to do this. In *Design* (November, 1960), Reyner Banham wrote of 'a sort of sporting relationship between the consumer and produce, of which market research is the outward symbol'. Production was no problem: the difficulty was to consume at a rate that would keep up with production. The value was now on the process rather than on the product; 'built-in obsolescence' was the answer. Entertainment, as Adorno remarked (1941: 26), is like a 'multiple-choice questionnaire' with no answers – the point is to choose *something*.

Money was circulating and culture was moving into the orbit of the masses. Communications technology, expanded by wartime research, moved ahead in leaps and bounds straight into the living rooms and routines of the nation. The speed of information and the way in which it was acquired changed: there was a nation hungry for immediate enlightenment; there was a nation eager for entertainment. Transistor radios blared, films flowed out from Hollywood. Everything was *fast:* fast food, fast cars, fast culture. Two cultures veered towards each other: the horror comic and the classic novel; the sex-pot and the sub-plot; the Hollywood musical and *Hiroshima Mon Amour;* Jerry Lewis and Jacques Tati; Norman Rockwell and Jackson Pollock. The news-stands told their own story: a new magazine, *Playboy,* offered naked women for fifty cents. The first fold-out featured Marilyn Monroe in her legendary calendar pose, with nothing on but the radio and Chanel Number 5. To see 'Woman' read women: soft-porn presented the process as the product – fast food for hungry eyes, cheap thrills for low budgets.

Within this consumer culture the body is proclaimed as a vehicle of pleasure: it is desirable and desiring and the closer the actual body approximates to the idealized images of youth, health, fitness and beauty, the higher its exchange value. 'Marilyn is not for sale', said her studio. This body-image circulates as a commodity in a socio-economic system, which thus further overdetermines the meanings represented. Advertising, feature articles and advice columns in magazines and newspapers ask individuals to assume self-responsibility for the way they look. Women are encouraged to see themselves in partial images, sexual segments: eyes, nose or mouth in the powder compact mirror, breasts, buttocks and legs in the media. The perception of the body within consumer culture is dominated by the existence of a vast array of visual images. Indeed, the inner logic of this culture depends upon the cultivation of an insatiable appetite to ingest images. The production of images to stimulate sales on a societal level is echoed by the individual production of images through photography (Sontag, 1978).

Monroe spoke of learning to 'reflect' herself through photographs. Christopher Lasch (1979: 47) has noted the profound effects of photography on the perceptions of social life:

Cameras and recording machines not only transcribe experience but alter its quality, giving to much of modern life the character of an enormous echo chamber, a hall of mirrors. Life presents itself as a succession of images of electronic signals, of impressions recorded and reproduced by means of photography, motion pictures, television and sophisticated recording devices. Modern life is so thoroughly mediated by electronic images that we cannot help responding to others as if their actions – and our own – were being recorded and simultaneously transmitted to an unseen audience or stored up for close scrutiny at some later time.

Day-to-day awareness of the current state of one's appearance is sharpened by comparison, with one's own past photographic images as well as with the idealized images of the human body which proliferate in advertising and the visual media. Women are most clearly trapped in the self-surveillance world of images; for, apart from being accorded the major responsibility in organizing the purchase and consumption of commodities, their bodies are used symbolically in advertisements.

The Hollywood cinema helped to create new standards of appearance and bodily presentation, bringing home to a mass audience the importance of 'looking good' and 'feeling fine'. Hollywood publicized the new consumer culture values and projected images of the glamorous celebrity lifestyles to a worldwide audience, giving it (literally) 'something to look up to'. The major studios carefully disciplined and packaged their film stars for audience consumption. To ensure that the stars conformed with the ideals of physical perfection, new kinds of make-up, hair care, and techniques such as electrolysis, cosmetic surgery and toupees were created to remove imperfections.

The 1920s, the period in which Norma Jeane Baker was born and brought up, was a crucial decade in the formulation of the new bodily ideal. By the 1930s, women – under the combined impact of the cosmetic, fashion and advertising industries, and Hollywood – had for the first time in large numbers put on rouge and lipstick, taken to short skirts, rayon stockings, and had abandoned the corset for rubber 'weight-reducing' girdles. In the orphanage, Norma Jeane was once given the 'special' treat of having her face made-up with powder and lipstick (see Monroe 1974). Such is the predominance of masculine mores, *maquillage* became a signal of women's readiness to conform to convention. Ever since its inception, the publicity machine of Hollywood has catered for and generated a great deal of interest in the 'backstage' areas, the private lives of the stars, their beauty tips, exercises and diet regimes. The beauty salon became a popular meeting place for mutual encouragement. The Hollywood fan magazines of the twenties and thirties 'indoctrinated their true believers with the notions that women are beautiful, men were manly, crime didn't pay, lovers lived happily ever after time after time, and Lana Turner was discovered eating a sundae at Schwab's Drug Store' (Levin 1976: 7). Pin-ups were big business during World War II, and remained so into the fifties. Theweleit (1987: 370) notes, 'The spectrum of stars and cover girls omits almost nothing; in the end, the same desirous gaze is trained on flesh-and-blood women as was trained on the image. Above all, they are desired through the *look.*'

By 1952 in the USA the cosmetics industry was selling over $1 billion worth of beauty products per annum. In 1948, 4,500,000 'breast pads' or falsies were sold as a result of the Hollywood 'sweater girl' trend. By 1950, 85 per cent of American women over fifteen wore bras, girdles, or both. Corsets had become a $500,000,000 annual business (see Ryan 1975). Dior's 'New Look' of 1947 encouraged a 'shift' of the erogenous zone: it emphasized the bosom, together with other feminine curves, so it was not surprising that women grew more conscious of the shape and size of the breast. By the fifties, the Hollywood model of small waist, uplift bra and full-lipped skirts made every woman into every man's desirable sex object. An article by Fay Vickers in *Photo World* of August 1946, 'The Great American Pin-Up', said: 'This form of art has found its way from the fox-hole to the home. Psychologists say there is no harm in it.'

Magazines such as *Photoplay, Silver Screen, Screen Book, Modern Screen* and *Motion Picture,* as well as publicizing the 'Secrets of the Stars' also offered readers the chance of self-improvement with advertisements claiming to provide remedies for over-sized busts, under-sized busts, fatness, thinness, and acne. Numerous kinds of bust-improvers were advertised, such as 'pneumatic busts' which could be inflated at will. The average bra size was a 34B, but fashion-conscious women padded out to Monroe's 38 or Jane Russell's 40. Publicity stills of the stars were retouched to eliminate any blemishes in the actor's appearance. In fact, Hollywood stars themselves began to rely less on aids and supporters to effect a given appearance, rather they carefully achieved the appearance of the 'body natural'. Body supporters such as the corset found less advocates in a culture which endorsed the exposure of the body on the beach and the wearing of casual leisure clothing.

This conscious 'body-work' continued to carry Hollywood overtones into the eighties: Elaine Deed writes in the *Guardian* (3 April 1986) of 'the general desire to achieve a stronger, more athletic figure – more like Jamie Lee Curtis than Marilyn Monroe, whose pin-up curves would look as out of place in today's gym as Twiggy among the voluptuously rippling flesh of The Sirens by Rubens.' Monroe had a pleasure – quite distinct from narcissism – in her own physique and her presence. 'I'm very definitely a woman and I enjoy it.' She wore her dresses one or two sizes too small, so that she was always conscious, from their clinging, of every part of her physique – 'To wear a dress like that,' says an admiring rival in *Niagara,* 'you've got to start laying plans at about thirteen.' It seemed that her exhilaration in her being and her body exhilarated *every* part of her; in *The Misfits* Arthur Miller has Monroe's character say, 'You can hear your skin against your clothes' – a remarkable articulation of pure sensuousness. Her appearance suggested inexhaustible energy and effervesence. One of the significant features of Monroe's image was the deviation from contemporary ideals of the 'sex symbol' – her refusal to wear a corset or girdle made her distinctive from the Hollywood woman's 'washboard stomach'. As her career progressed her physical problems further diverted her image from the stereotype. Monroe transgressed a number of unwritten celebrity sex symbol conventions by making fun of her appearance: she gave reporters her 'upper' and 'lower' hip measurements, and consented to being photographed as a fairly comic figure. She made noticeable what once seemed natural: the little voice, the walk, the wiggle, the 'womanly' shape. In 1953, Monroe lay down on the sidewalk outside Grauman's Chinese Theatre on Hollywood Boulevard; beside her was Jane Russell, her co-star from *Gentlemen Prefer Blondes.* They placed their hands in wet cement and scrawled their names alongside the prints. Monroe, acknowledging the fetishization of their bodies, suggested that Russell (for whom Howard Hughes had designed a special brassiere) should bend over the cement; as for herself, Monroe proposed sitting in it.

The image-repertoire of popular culture encourages the body to project an image of itself. The sights that daily assault the individual in consumer culture do not merely serve to stimulate false needs forced on to the individual. Part of the strength of consumer culture comes from its ability to harness and channel genuine bodily needs and desires, albeit that it presents them within a form that makes their realization dubious. The desire for sexual fulfilment, youth and beauty represents a reified entrapment of the trans-historical human longings within distorted forms. However, in a time of diminishing economic growth, permanent inflation and shortages of raw materials, the contradiction within the cultural values become more blatant – for the old, the unemployed, the disabled, low-paid – but also for the privileged few who participate most actively and experience more directly the chasm between the ideal of the imagery and the exigencies of everyday life. It is quite extraordinary how frequently journalists wrote critical articles about Monroe 'letting herself go' when they were well aware of the acute physical disorders she was suffering from.

In *Seeing Through Clothes* (1978) Anne Hollander notes that the human body as portrayed in painting and sculpture changes its shape to fit the fashions of the time; that 'all nudes in art since modern fashion began are wearing the ghosts of absent clothes – sometimes highly visible ghosts.' Photography, rather than liberating our perception of the body, has helped to tie it closer to fashion. Through a biased choice of models and poses it seems to offer scientific proof that we are – or ought to be – the 'right shape' for contemporary clothes. When posing for photographers, late-Victorian nudes protruded their behinds like bustles; 1920s' nudes adopted a debutante slouch, and nudes of the 1940s tucked in their tummies and hips and stuck out their chests to produce the 'flat-bottomed, melon-breasted' figure then considered most desirable. The one-piece bathing costume makes the body conform to the shape of the costume, not vice-versa. The anonymity of the image is reinforced by the poses, conventional and common, which also leave the body unprotected and apparently 'available'. The conventions of glamour photography work to construct the body as spectacle: the female body is a spectacle because parts of it – the parts that say 'this is a woman' – are deemed pleasurable to look at.

According to Wilson (1985: 157), 'It was above all the camera that created a new way of seeing and a new style of beauty for women in the twentieth century. The love affair of black and white photography with fashion *is* the modernist sensibility.' Billy Wilder said of Marilyn Monroe: 'The first day a photographer took a picture of her, she was a genius' (in Zolotow 1961: 261). In her early stills, Monroe is seen as interesting because of her body; in later portraits, it is a 'celebrity body' that is the interesting sight. In the silent movies stylization of both gesture and looks was necessary for narrative, and prompted not only new ways of walking, sitting and using the hands, but also the development of styles to suit personalities. Fashions became part of a mammoth tie-up between the cinema and big business; the two influenced each other in the interests of the 'image industry'. Glamour contains a charm enhanced by means of illusion. A glamour image of a woman is particularly impressive in that it plays on the desire of the viewer in a peculiarly pristine way: beauty or sexuality is desirable exactly to the extent that it is idealized and unattainable. Monroe suffered because of this image: 'People expected so much of me, I sometimes hated them. It was too much of a strain. . . Marilyn Monroe has to look a certain way – be *beautiful* – and act in a certain way, be talented. I wondered if I could live up to their expectations.' (In Weatherby 1976: 146)

The desire for the unattainable 'glamour girl' reflects only too accurately the real unattainability of the woman represented, and also of the image of perfection glamour pictures offer. The perfectly presentable women in the glossy advertisements project a vision of perfection few women can ever attain. The desire for such exalted standards is displaced on to desire for the products they advertise or connote, the admirers they seem to effortlessly attract. For the model, the posing offers space to merchandise a self; for the photographer, the posing offers the promise of a saleable 'signed' self. As Goffman (1985: 18) has argued, the glamour portrait 'is already a rekeying, already a ritualization of the human form, already a departure from the simple rendering of an aspect of the world the way it is for us'. Women's bodies and exchange became identified: representations of women became the commodities that movie producers were able to exchange in return for money.

Glamour portraits are 'made-up': not only in the simple fact that cosmetics have covered the body, but also because the images (rather than the women) are constructed and contextualized. Goffman thus notes that 'the possibility of arranging a scene from the visual pinpoint of view of a single camera's eye – into which angle and distance of vision vast hordes of viewers can be thrust – is a social license as well as an optical one'. The photograph is not a 'window on the world' nor a pathway to the presence of the author; it is a 'screen', shielding and showing a social relationship. The photographs of Marilyn Monroe, as well as possessing exchange-value in particular markets, also circulate a currency of codes through which 'woman' is constructed in representation. Solitary or not, the figures in the stills implicitly address themselves to us, the spectator, position us nearby through our being allowed to glimpse what we can glimpse of them, thus stimulating a social situation of sorts.

Images of women have traditionally been the province and the property of men. Monroe remarked: 'When the photographers come, it's like looking in a mirror. They think they arrange me to suit themselves, but I use them to put over myself. It's necessary in the movie business, but I often hate it. I never show it, though. It could ruin me. I need their goodwill. I'm not stupid' (in Weatherby 1976: 154). She also said, 'In Hollywood a girl's virtue is much less important than her hair-do. You're judged by how you look, not by what you are.' In 1949, photographer Tom Kelley had a special assignment. He asked Monroe, a starlet struggling to stay in contention at the studios, if she would pose nude for a 'tasteful' calendar shot. When she asked what it would pay and was told 'fifty dollars', she agreed. Kelley and his wife, Natalie, were convinced that Monroe was a brilliant model in sensing what the camera saw as she posed on a red velvet drape. The calendar came close to being banned from the mail; the post office considered it pornographic. When Kelley heard of the calendar company's dilemma, he had the better of the two final shots analysed by a committee of well-known artists, whose considered judgement was that it was a work of art 'in perfect symmetry'. The company went on to make $750,000, Kelley earned $900, and Monroe was given a $50 fee. When the identity of this 'Miss Golden Dreams' was later revealed Marilyn Monroe explained her reasons for the photograph. The papers said, 'the public can forgive a delicious but empty-headed blonde almost anything.'

Mechanical reproduction of still photographic images offered a breakthrough in the public availability of pornography and glamour portraits. The advent of cheap, mass-produced visual erotica opened the market to the masses. Magazines, books and posters provided a context which does not demand literacy in any language and is highly flexible – it can be used in numerous social circumstances and locations: Monroe's famous 'skirt' pose from *The Seven-Year Itch* has been exhibited on a print, a poster, a postcard, a painting, a billboard, a matchbox and a button-badge. Photography does not simply reproduce the real, it recycles it – a key procedure of modern society. This recycling makes cliches out of unique objects, distinctive artefacts out of cliches. Images of real things are interlayered with images of images.

One of the defining features of photography as against certain other forms of visual representation is its capacity to appear 'truthful'. A human artist may filter the 'real world' through the creative imagination, but the camera and lens are often regarded simply as pieces of machinery that enable an image, a duplicate, of the world to be transferred on to film. A photograph projects the 'evidence' that whatever is within the frame around the image 'really happened'. However, photography actually involves just as much artifice as does any other mode of visual representation (Barthes, 1981; Boorstin, 1962). The reality of the scene is selective, optional, fantastic. A photograph does not present us with 'likenesses' of things; it presents us, we can believe, with the things themselves. This belief clearly makes us ontologically restless: obviously a photograph of Marilyn Monroe is not Marilyn Monroe in the flesh. However, it sounds equally odd to hold up a photograph of Monroe and say, 'This is not Marilyn Monroe.' Such confusions suggest that we do not actually know how to place the photograph ontologically.

The problem is not that photographs are not visual copies of objects, or that objects cannot be visually copied. The problem for us is that even if a photograph was a duplicate of an object, it would not bear the relation to its object that a recording bears to the sound it duplicates. Such publications as *Marilyn Monroe: Her Life in Pictures* are actually involved in an eminently conceited project. A record reproduces its sound, but a photograph does not reproduce a sight. It seems that objects are too 'close' to their sights to release them for reproducing; in order to reproduce the sights they make, you have to reproduce *them* – take an impression, make contact.

Photographs certainly do have *pretensions* to reality: they say to the spectator, this is how it is (you need look no further), you have only to recognize this record. At one extreme, news photography and social realist photography are boldest in their claim to 'tell it like it is': Anthony Summers' *'Goddess* uses a cover illustration to compliment the 'sober' rhetoric of the text – a black and white press photograph of Monroe after her divorce from DiMaggio, looking tired and anguished, 'caught' by the camera. On the other hand, 'art photographs', bearing the signature of an *'auteur'*, freely admit there has been creative intervention on the part of the artist-photographer, (such as Milton Greene), so that claims to authenticity in relation to a 'real world' become attenuated. In general, photographs connote authenticity when what is seen by the camera eye appears to be an adequate stand-in for what is seen by the human eye. Photographs are coded, but usually so as to seem uncoded. A fact of considerable social importance is that the photograph is a *structuring space* wherein the reader deploys, and is deployed by, conventional codes in order to *make* sense. The medium is one of the signifying systems that produce the ideological subject in the same movement in which they 'communicate' their ostensible 'contents'. Photography draws on an ideology of the visible as the veritable, of 'seeing is believing'. To complete the circuit of recording, visibility and authenticity established by the photograph, there must be someone looking at it. The spectator looks at Monroe's image, and the look of the lens is completed by the look of the spectator: the photograph suggests that these two looks are alike. Meaning is produced, finally, in the spectator's look.

The spectator's look at a picture of Monroe is not limited in time: she or he may merely glance at the image, study it at length, stand at a distance or close to, pass by it or return to it time and again. It is not unusual to look at a photographic still in isolation: it may be in a magazine, within a television programme, in a newspaper surrounded by print or captioned, or it may be one of a series of images intended to be read in sequence as a narrative. The immediate context in which an image appears will set limits on the ways in which it is likely to be read. The spectator's look is thus the key in the reading of the image. The look is inflected in specific ways in relation to photographic stills as against, say, motion pictures. The spectator can choose to gaze at indeterminate length in the presence of the photograph, treat it with reverence or copy it (Warhol), in *Montage* (Peter Blake), or cut-up (James Rosenquist). Looking may lead to contemplation, sometimes to voyeurism: 'Scopophilia' contributes to the 'drive to master', a component of sexually based pleasure in looking. (Freud illustrates 'The Uncanny' with a reference to the take of the living doll Olympia, see Hoffmann 1953–74, vol. 17: 219–52.) A voyeur's pleasure hinges on the object of the look being oblivious to the actual observer: to this extent, it is a pleasure of power, and the look is a controlling one. Photographs are eminently well-equipped to elicit this pleasurable feeling: the apparent authenticity of what is in the image colludes with the fact that it is obviously not (all) there – and so can be looked at interminably, for the circuit of pleasure will never be broken by a renewed look.

Billy Wilder was more aware than most of Monroe's special photogenic qualities:

> Every movie star has a certain voltage. It is as if one were to hold up a
> light meter to the screen and certain stars will register more than
> others. Well, Marilyn has the highest voltage. . . She never flattens out
> on the screen. In any scene she is in, there is never what I call a hole in
> the screen. She never gets lost up there. You can't take your eyes away
> from her. You can't watch any other performer when she's playing a
> scene with somebody else. (In Zolotow 1961: 201)

The close-up not only encourages a sense of intimacy, but also invites us to read character into the face, to formulate, from the lines and wrinkles, an impression easily mistaken for a detailed and knowledgeable portrait of the star. Monroe suffered from a nervousness which never settled into shyness. In her movie appearances there is sometimes a strain, in the editing and in the actors, in her face, in her speech, that tells us of the communal anxiety as to whether she will remember her lines and realize when and where to move. In contrast, in stills Monroe consistently shows authority and pleasure: she could simply 'be', seeming poetically present. As Bert Stern (1982: 54) comments: 'She'd move into an idea. I'd see it, quickly lock it in, click it.' The photographers were her escorts to the public, the visionaries who recognized she was in the realm of aesthetics whenever she stopped still.

Monroe's clothes designer, Billy Travilla, recalls her telling him:

> 'I can make my face do anything, same as you can take a white board
> and build from that and make a painting.' But the only way she was
> highly sexed was the charge she got out of looking in the mirror and
> seeing that beautiful mouth that she'd painted with about five shades
> of lipstick, to get the right curves, the right shadows to bring out the
> lips, because her lips were really very flat. (In Summers 1985: 83)

Monroe's one-time partner and photographer, Milton H. Greene, adds that

> It costs a lot of money to look like Marilyn Monroe. You don't just wake up in the morning and wash your face and comb your hair and go out in the morning and look like Marilyn Monroe. She knows every trick of the beauty trade. She knows what has to be done to make her look the way she wants to look. (In Zolotow 1961: 194)

Yet Monroe's image is not the sum-total of various 'tricks' and photographic techniques. The work concerning the social construction of the body has been extremely illuminating, but it would be a specious account that claimed that beneath the camera, colour and cosmetics there is 'just another' person. Such an argument is far too cynical and simplistic. Monroe co-starred in *Niagara* with Joseph Cotten. He said at the time:

> A lot of people – the ones who haven't met Marilyn – will tell you it's all publicity. That's malarkey. They've tried to give a hundred other girls the same publicity build-up. It didn't take with them. (In Zolotow 1961: 128)

Emmeline Snively, who founded the Blue Book Model School, makes a similar point:

> Girls ask me all the time how they can be like Marilyn Monroe. And I tell them, if they showed one tenth of the hard work and gumption that that girl had, they'd be on their way. But there will never be another like her. (In Taylor 1984)

Monroe possessed a particular quality which, when photographed, was complimented by the particular qualities of the medium. What she came to know and to perfect was how to exploit and to enhance her natural abilities. Richard Avedon has said:

> She understood photography, and she understood what makes a great photograph – not the technique, but the content . . . she was more comfortable in front of a camera than away from it . . . she was completely creative . . . she was very, very involved with the meaning of what she was doing, in an effort to make it more, to get the most out of it. (In Taylor 1984)

Earl Theisen reflected on this theme after photographing Monroe:

> Everything she does for a camera has been studied carefully. She knows exactly what she's doing. You can watch her, as you're focusing

your camera on her, getting ready to turn it on. She knows exactly
how far she wants to open her mouth, how much to raise her upper
lip. (In Zolotow 1961: 113)

The still photographers' response to Monroe, as a person, seems to have
been directly opposite to that of her movie directors. Where the latter saw
insecurity, inconsistency and panic, the former were aroused to passionate awe – sending otherwise taciturn people such as *Vogue* photographer Bert Stern (1982) to hyperbole:

She was the wind, that comet shape that Blake draws blowing around
a sacred figure. She was the light, and the goddess, and the moon. The
space and the dream, the mystery and the danger. But everything else
all together too, including Hollywood, and the girl next door that
every guy wants to marry. I could have hung up the camera, run off
with her, lived happily ever after.

Forever frustrated by seeming simple-minded in movies, in stills Monroe
exudes intelligence. The person looking at stills of Monroe seems to hear
her private voice, a sound which bestows a fine knowingness on her, a
thorough command of the medium. She commands us from the still as she
cannot from the screen.

As an image, Monroe hides from reality; as a human being, she haunts
reality. Photography maintains the presentness of the world by accepting
our absence from it. The reality of a photograph is present to us while we
are not present to us; a world we know, and see, but to which we are
nonetheless absent, is a world past. The camera, being finite, crops a
portion from an indefinitely larger field: continuous portions of that field
could be included in the photograph taken; in principle, it could all be
imprinted. Objects in photographs are aimed at, shot, stopped dead in
their tracks. The image thus displays the *absence of the absence* – what
has gone seems to remain. The implied presence of the rest of the world,
and its explicit rejection, are as essential in the experience of a photograph
as what it explicitly presents.

'Monroe' became a modern adjective in the 1960s: blonde women stars
were 'Monroe-like', 'Monroe-isms' were identified, and her name was
invoked when the famous met tragic ends. Monroe died early in the
sixties, and the flow of photographic images was thus stemmed and
seemed finite. As Monroe was being carried into the mortuary at
Westwood Memorial Park, a swarm of photographers ran over graves
and flowers, desperate to frame the scene, eager for one parting picture.
Any lingering sense of decency was obscured by each photographer's
desire to be a winner in the competition for images. Photography may

function as a 'collage machine': a camera selects and transfers a fragment of the visual continuum into a new frame. In photomontage, the photographic images are themselves cut out and pasted into new, surprising, provoking juxtapositions. When the images of Monroe seemed suddenly in short supply, the artist became *bricoleur*, recontextualizing objects in an effort to produce novel insights and ideas.

'Once you "got" Pop', wrote Andy Warhol, 'you could never see a sign the same way again.' The Combines, Assemblages, Environments and Happenings of the late fifties and early sixties manifested the 'collage' principle, imparting a freshness into the hermetic art academy. The sixties witnessed a celebration of the shallow, with Warhol calling his studio 'the Factory' and manipulating his 'employees' in the manner of an old-time movie mogul. Monroe became his Muse – his amusement. 'Low-Pop' and 'Fine-Pop', moptops, Op Art and Pop Art, having their kitsch and treating it, the 'new' culture catered to the possibilities of what Charles Taylor (1985) called 'affluence-in-privacy', the modern mode of consumption. The acceleration and increase in stylistic cross-fertilization, coincident with the plugging-in of 'the electronic global village', meant the easier transcendence of national boundaries, and the emergence of a modern 'international style' of visual communication. Greater affluence meant more public entertainment: the number of prestigious international exhibitions and festivals increased. Making the waste-products of a society into new commodities, a trend began for chi-chi junk and repackaged nostalgia. The symbols of Marilyn Monroe and James Dean as stars who died young were soon joined by the 'martyrs' of pop psychedelia: 'Hope I die before I get old' was a popular theme of the time.

What Pop Art did was to banish the heroic necessity and to stylize the machine-made mediocrity of commercialized culture, by aestheticizing it and sometimes mildly satirizing it. The public, looking at the image offered by Pop Art, found itself in the position of the Bourgeois Gentleman, delighted to discover he had been speaking prose all his life without knowing it. Now the public was informed that all the while their eyes had been sliding across comic books, TV screens, move-star posters, even the American flag, they had been appreciating art without realizing it. This new art was, in itself, infinitely reproducible in a way that no art prior to it ever had been, since it drew on material that was in itself originally designed to catch the eye with a bold, simple flourish or to be conveniently duplicated by even the crudest technology. One of the principal effects of Pop Art in general, and of Warhol's work in particular, was to popularize the idea of 'multiples' – not only the repetition of a single image many times in a single work, but also like the art forms (such as posters and prints) which 'democratized' the artist's original concept.

In August 1962, Warhol gave up the rubber-stamp method he had been using to repeat images and took up a silkscreening process, whereby photographs are blown up, transferred in glue on to silk, and then rolled across with ink so that the ink seeps through the silk but not the glue. Warhol's first experiments were with heads of Troy Donahue and Warren Beatty; then, when Marilyn Monroe died in that same month, he made the first screens of her face.

Walter Benjamin argued that what 'withers in the age of mechanical reproduction is the aura of the work of art'. By this he meant that the peculiar singularity of any given object arises from the fact that it is created at a particular point in time and space, in a response to a particular set of circumstances. However, 'the technique of reproduction detaches the reproduced object from the domain of tradition,' pries it from its frame, thereby depriving it of its 'aura'. Consequently, the beholder of the reproduced object is moved to incorporate it into her or his own psychological domain. The difference between seeing a picture of Monroe in the formal, historically aware context of the museum and seeing it reproduced on your living-room wall or in the magazine on your lap is similar to the difference between hearing a symphony in a concert hall and playing a recording, seeing a movie in a theatre and watching it on video. Denuded of proper context, and of a sense of auspiciousness, these secondary images serve 'the desire of contemporary masses to bring things "closer" spatially and humanly, which is just as ardent as their bent towards overcoming the uniqueness of every reality,' as Benjamin remarked. Campbell's soup cans or Brillo boxes may seem faintly comic in their invocation of the quotidian, but as presented on a Warhol canvas, their images multiplied and magnified, they also remind us of where the true power in the West lies – in mass production and in the massive enterprises required to create and sustain such production. So it is with Warhol's special celebrity series – the Elvises, Lizs and Marilyns: they remind us that celebrity power as we now read it arises precisely from a similar power – to endlessly replicate a product and pester the public with it until it is accepted.

Despite the nagging newness of consumer culture, the pulse of the then recent past was still felt in the 1960s and 1970s with extraordinary intensity. Paris in the fifties and sixties, belatedly exposed to the 'classic Hollywood' movies, lit the fuse on the trail of nostalgia that smouldered and sparked through to the 1980s. *Cahiers du Cinema* fostered a new reverence for the Hollywood director – such as Howard Hawkes and Alfred Hitchcock. Well-worn reputations experienced a sudden face-lift. With graphics returning to posters and Pop Art becoming 'Pop*ism*', the airbrush was in favour in commercial art. It portrayed people with no trace of any blemish – flawless and idealized products. Posters became the leading popular art form of the sixties and early seventies, a vehicle for psychedelic extravaganzas (Monterey in 1967) or insurrection (Paris in 1968). Public interest in these enlarged icons was shown by the widespread sale of large poster reproductions by firms such as Athena: the two chief sellers were images of Che Guevara and Marilyn Monroe. The poster 'Marilyn' seemed like E. M. Forster's paradoxical picture of America: 'gigantic and homely'. The 1970s reaffirmed the re-cognition of the recent past: universities and polytechnics began teaching 'film studies' courses, and museums launched exhibitions of fifties' art. Literature on popular figures came to warrant libraries of its own. Jane Dorner's *Fashion in the Forties and Fifties* appeared in 1975; Alan Jenkins' *The Forties* in 1977; and Peter Lewis' *The 1950s* in 1977. George Lucas stylized the 1950s in *American Graffiti* (1973). The stage and screen versions of *Grease* and the film *That'll be the Day* conveyed the pop attractions of the period to the generation born from 1955 onwards. The books on Monroe by Norman Mailer ensured that her myth remained in public prominence. The death of Elvis Presley in 1977 brought another surge of fifties nostalgia. The large auction rooms moved into forties and fifties 'antiques'. Television screened a regular flow of old motion pictures. The recovery of rare 'Marilyn' images was increasingly urgent: the compiler of picture books reminded one of Steve McQueen in *The Great Escape*, speeding towards freedom, with the precious prints now safely in his hands. In 1986, the publication of another book of photographs – André de Dienes' *Marilyn Mon Amour* – merely marked a further moment in the pictorial post-history of Marilyn Monroe, a moving projection of the past into the present, a mechanical death anticipating a spiritual resurrection.

The British actress Constance Collier, who late in her life came to teach Marilyn Monroe, said to Truman Capote (1981: 207–8) of her pupil:

> What she has – this presence, this luminosity, this flickering intelligence – could never surface on the stage. It's so fragile and subtle, it can only be caught by the camera. It's like a hummingbird in flight; only a camera can freeze the poetry of it. But anyone who thinks this girl is simply another Harlow or harlot or whatever is *mad*. . . I hope, I really pray, that she survives long enough to free the strange lovely talent that's wandering through her like a jailed spirit.

Contrary to the opinion of those authors who tell us that Monroe 'is still with us, in films and photographs', one can learn more about oneself and one's contemporaries when one contemplates the image of Monroe. The present winds that past around itself like the tape is wound on the spool of a recorder. Mailer (1973: 15) has neatly reviewed the aftermath of Monroe's death,

> as Jack Kennedy was killed, and Bobby, and Martin Luther King, as Jackie Kennedy married Aristotle Onassis and Teddy Kennedy went off the bridge at Chappaquiddick, so the decade that began with Hemingway as the monarch of American arts ended with Andy Warhol as its regent, and the ghost of Marilyn's death gave a lavender edge to that dramatic American design of the Sixties which seemed in retrospect to have done nothing as to bring Richard Nixon to the threshold of imperial power.

As Sammy Davis Jr admitted, 'Still [Monroe] hangs like a bat in the heads of the men who knew her.'

'I failed at first to recognise her,' says Proust's Marcel, in *Remembrance of Things Past*, vol. 3 (1983: 990), upon seeing the ageless Odette after a lapse of many years, 'not because she had but because she had not changed.' Walter Benjamin (1982: 100) wrote that a man who dies at the age of thirty-five is remembered 'at every point of his life [as] a man who dies at the age of thirty-five' (1982: 100). Thus with a woman who dies at the age of thirty-six, we cannot divest knowledge of her subsequent demise from our photographic memories of her earlier years. The prime function of memory is not to preserve the past but to adapt it so as to enrich and manipulate the present: 'The remembered image is combined with a moment in the present affording a view of the same object,' Roger Shattuck (1964: 47) remarks, 'Like our eyes, our memories must see double; those two images then converge in our minds into a single heightened reality.' Monroe's image changes according to the context wherein it is seen: her photograph seems different on a contact sheet, in an FBI file, in a fan magazine, through a shop window, on a billboard, on a bedroom wall. Each of these situations suggests a different use for the photographs but none can secure their meaning. It is in this way that the presence and proliferation of these photographs contributes to the revision of the very notion of meaning, to that parcelling out of the truth into relative truths which is taken for granted by modern liberal consciousness.

Men have made the image of Marilyn Monroe, and they have often underwritten that image. Benjamin believed that the right caption beneath a picture could rescue it from the ravages of modishness and liberate its utility. He urged writers to begin taking photographs. Concerned writers have not often taken to cameras, but they are frequently enlisted to spell out the truths to which photographs testify. Captions may speak louder than pictures, but no caption can completely capture the picture's meaning. In a consumer culture, even the well-intentioned and wisely captioned work of photographers issues in the discovery of beauty. Cameras miniaturize experience, transform history into spectacle. Photography's 'realism' creates a confusion about the real and the unreal which is morally analgesic as well as sensorily stimulating. It clears our eyes, yet it may well blur our memories; as Barthes (1981: 92) put it, '*Life/Death:* the paradigm is reduced to a simple click, the one separating the initial pose from the final print.' It is not reality that photographs make immediately accessible, but images; they are not so much an instrument of memory as an *invention* of it.

Whatever the shortcomings of capitalism's ability to delivery consumer goods and the consumer lifestyle to all sectors of the population, it has never been short of images – and for those who inhabit the negative side of consumer culture, consumption is limited to the consumption of images. The Russian journal *Nadya* once commented, 'When we think of the American way of life, we think of bubble-gum, coca-cola, and Marilyn Monroe.' It is ironic that the most public of people could experience such solitude: too many cameras and not enough care. The man who 're-beautified' Marilyn Monroe after her autopsy was Allan 'Whitey' Snyder, her regular make-up artist, and in his pocket was a gold-plated money-clip she had given him years before, with the inscription: 'To Whitey. While I'm still warm. Marilyn.' 'She looked beautiful,' a mourner said after the funeral, 'like a beautiful doll' (*New York Herald Tribune* 9 August 1962).

Monroe once wrote (1974: 135): 'I want to be an artist, not an erotic freak. I don't want to be sold to the public as a celluloid aphrodisical. Look at me and start shaking. It was all right for the first few years. But now it's different.' Monroe had a surfeit of all those qualities which myth and the movies instruct us that 'all blondes ought to have' – but only she had them and appreciated them and utilized them in such a distinctive, moving and memorable way. Her photographic image captures her at her most assured and assertive; the conspiratorial laugh is inviting the spectator to share in her subversive style. America's most visible blondes, Jean Harlow and Marilyn Monroe, experienced the most pathetic deaths. They were shown as shallow and celebrated as such – good-time girls (as long as they lasted) of no identifiable background, no parentage and no pedigree except for American blondeness and the egalitarian dream. In focus, Monroe was both the object and the originator of power: she was the locus of the masculine look, and she signified so much more than the 'symbol' she subverted. It is the absence today of the person that makes the ubiquity of her image so poignant.

Some Like It Hot: 'I tell you it's a whole different sex!'
Photograph courtesy of The Museum of Modern Art Film Stills Archive.

4

Marilyn in Movies

He who has once begun to open the fan of memory never comes to the end of its segments; no image satisfies him, for he has seen that it can be unfolded, and only in its folds does the truth reside.

Walter Benjamin

I'm trying to find myself as a person, sometimes that's not easy to do. Millions of people live their entire lives without finding themselves. But it is something I must do. The best way for me to find myself as a person is to prove to myself that I'm an actress.

Marilyn Monroe

'They've tried to manufacture other Marilyn Monroes, but it wouldn't work. It never will, she was an original and there'll never be another.' Billy Wilder made the comment after Monroe's death, before the succession of BBs and DDs attempted to fill the 'movie queen' mantle vacated by 'MM'. In her final interview she predicted the trend: 'These girls who try to be me – I guess the studios put them up to it, or they get the ideas themselves. But . . . they haven't – you can make a lot of gags about it – like they haven't got the foreground or else they haven't got the background. But I mean the middle, where you live' (Meryman 1962). The contribution of movies to the myth of Marilyn Monroe has been manifold: by putting into circulation a whole series of Monroe lookalikes, by repeating the re-runs of old images, and by recalling the presence of the person, the movies have encouraged the kind of verdict expressed by Conway and Ricci (1964: 21) 'As long as there are movie projectors, future generations will be able to see this brilliant artist . . . a girl who was truly an American Phenomenon.' Can we still (did we ever) really 'see' Marilyn Monroe?

Contemporary viewers of Monroe hailed her as the 'ultimate movie sex goddess'. European film critics expressed respect for Monroe's acting abilities, and she received awards and accolades in both France and Italy. Anglo-American critics, in contrast, were never convinced that Monroe was anything more substantial or more significant than a highly successful sex symbol. Such an image was both a blessing and a burden in an era where the limits of acceptable movie erotica were steadily extended. Switched-on sex increased as the studios scrambled to out-sell television. The cinema showed renewed interest in exploiting the relationship between viewing and sexual arousal: voyeurism depends on the viewed occupying the same real space that shelters the viewer. The movie's ease in shifting place, transferring time, switching perspective, makes it a model of technological grace. In combining the various arts – music, drama, spectacle – the movie is also formally the anticipation of a community it alone will be capable of attending to. As we look up there, we lose ourselves here: our sudden impersonality facilitates our interest in the moving figures. In the cinema, an intensely male-dominated institution, the rule was the representation of women by the men for the men: for your pleasure.

The 'sex symbol' was the movie man's 'curious object of desire', a woman defined by the camera as interesting because of her gender. She serves as spectacle: ideally, the story should pause in order to leave the screen free to focus on the body, and when the woman does contribute to the action we watch her voyeuristically. When Monroe performs 'I Want to be Loved By You', there is a momentary wondering whether she will overflow into the camera; it owes something to the precision with which the director has the margin between spotlight and shadow fall on the most crucial part of the side area of her bosom. When we watch Monroe in movies, we watch with masculine vision and masculine values.

The medium of film would seem to contain the strongest claim to 'capturing' Marilyn Monroe. Film retrospectives boast 'Marilyn lives!' and film buffs cover the pull-out, pin-up, poster prints of 'MM'. Indubitably, the Monroe image celebrated, with considerable financial reward, female sexuality, and yet at the same time could not be allowed to undermine sexual mores in which a capitalist society is implicated. More detailed knowledge of what a star brings to a film in terms of meanings built up elsewhere strengthens our understanding of what meanings are in play in any individual film, and provides concrete possibilities for evaluating the interplay between the film on the screen and the wider institution of cinema, including the knowledge brought to the film by the audience. In order to grasp the significance of Monroe as a movie star, one must make some effort to understand the period and place from which she emerged, the medium in which she worked, and the personal qualities she brought to her performance. The following discussion thus concerns Monroe's relationship with Hollywood, how her image was used and with what individual effects.

The ideological, Barthes said (1986: 348), 'is the Cinema of a society'. Filmic images are not *hand*-made; they are manufactured. What is manufactured is an image of the world. The world of a moving picture is screened. What does the silver screen *screen*? It screens us from the world it holds – that is, makes us invisible. It also screens that world from us – that is, screens its existence from us. Monroe is *there*, on the screen, now, tomorrow, at the flick of a switch; and she is obviously elsewhere, and what we see is something else. The altering frame is the image of perfect attention. Early in its history the cinema discovered the possibility of *calling* attention to persons and parts of persons; but it is equally a possibility of the medium not to call attention to them but, rather to let the world 'happen', to let its parts draw attention to themselves according to their presumed natural weight. There is usually nothing 'accidental' about the appearances of Marilyn Monroe on the screen: the extraordinary degree of attention accorded to her person by the camera was largely determined by her sex and by her star status.

Laura Mulvey (1975) suggests that there is a tension in narrative films between the demands of narrative (forward movement – what is going to happen next?) and the demands of spectacle in the form of the woman (stasis – stop and look at her). For example, in the famous 'skirt' scene in *The Seven Year Itch*, the plot has drawn to a halt as we watch Monroe displaying her body. Because, Mulvey says, the image of woman is inherently threatening, the presentation of woman as pleasing spectacle has to defuse this threat – by fetishizing the woman (a mode of presentation wherein the woman's body or parts of the body, or else her clothing or cosmetics are rendered as things in themselves) or by punishing her (stories wherein the woman is shown as guilty and made to suffer at the hands of the hero, either by direct punishment or simply by being exposed to him). The cinematic rhetoric of lighting, colour, framing, composition, placing of actors (referred to as *mise-en-scène*) contributes to this way of seeing women: a position constituted in terms of male heterosexuality, a viewpoint both disrespectful and dominating. Monroe was certainly subjected to the techniques of cinematography, but she just as surely subjected them to a form of irony. Speaking of Mae West, Monroe said: 'I learned a few tricks from her – that impression of laughing at, or mocking, her own sexuality.'

After Monroe's first major movie, we see a new star, only distantly a person. 'Marilyn' means 'the figure created in a given set of films'. Her presence in these films is who she is, not merely in the sense in which a photograph of an event is that event; but in the sense that if these films did not exist, *Marilyn would not exist,* the name 'Marilyn' would not mean what it does. The figure it names is not only in our presence, we are in hers, in the only sense we could ever be. That is all the 'presence' this 'Marilyn' has. Through a succession of projections, light and shadow delineate a delicate form of familiar shapes and scenes. It is enlightening to recall those stars who have provided the movie camera with human subjects – individuals capable of filling its needs for individualities, whose inflections of demeanour and disposition were given full play in its projection. They provided staples for impersonators: one gesture or syllable of mood, two strides, a wink, a cough, a shrug of the shoulders, half a smile, or a passing mannerism was sufficient to single them out from all other creatures. They realized the myth of singularity. Monroe's singularity made her more 'like us' – or rather, made her difference from us less a matter of metaphysics, to which we must accede, than a matter of responsibility, to which we must bend. This made her even more glamorous: she could, it seemed, stand upon her singularity.

Monroe fascinated the movie world more than any other star since Garbo, to whom she was sometimes compared. Garbo worked mostly in tragedy and Monroe in comedy, and they were enormously different in temperament, but both could make a poor film worth watching. The camera found incandescent qualities in them both (what Billy Wilder terms 'flesh impact'), and in return they responded wholeheartedly to it; from both of them there was an extraordinary love directed towards the camera, to the leading man and to the audience. Both were deemed 'instinctive' actors. Joshua Logan thought Monroe 'the most completely realized and authentic film actress since Garbo. Monroe is pure cinema. Watch her work. . . How rarely she had to use words. How much she does with her eyes, her lips, her slight, almost accidental gestures.' Monroe was not merely 'a woman in movies' – many preceded her, many more followed her: she was an exception in what Wilder called 'a celluloid Detroit, a factory town for mass production out of which something good comes now and again'.

Graham Greene's description of cinema as a 'womb with a view' is perhaps too naturalistic a metaphor to capture the atmosphere behind the complex wiring of an elaborate modern medium. John Huston, a puckish figure in a risk-aversive environment, chose to call Hollywood 'a sort of cage':

Stay in it too long and you develop an institutional neurosis. The only way to avoid this is to prowl around on the outside of the cage and

see it from a wider perspective, then fight for whatever values you
have. . . In the long run, it's up to you to make whatever you can of it.
Our business is to capture dreams. We press the dream on celluloid
and run it to audiences in darkened cinemas. We sit in the dark womb
of the auditorium and share the dream. . . Hollywood has always been
a cage, a cage to catch our dreams. (In Hamblett 1966: 174–5)

The critic Greene and the director Huston both entertain the idea of an
immaculate conception; both, however, are well aware of the material
meddling with this conception – which is, perhaps, why they so
steadfastly cling to such mystical images.

Movies rely on our experience of other movies. 'I thought they burned
that,' says Groucho Marx to the camera, looking at a child's sled called
'Rosebud'. 'Haven't we met someplace before?' Cary Grant asks Ralph
Bellamy in *His Girl Friday* (1940); they had, three years before, in *The
Awful Truth*. Such allusions in Hollywood movies refer openly to what
we all know: that the movies are a 'projected world', a mythology made
up of a limited number of stories and figures. As Dorothy Parker
remarked. 'The only "ism" in which Hollywood believes is plagiarism.'
The only plagiarism praised by Hollywood was of the profitable kind:
'classic' genres, well-known types, well-loved themes. The 'classic' period
for Hollywood, roughly 1930–45, directly preceded the rise of Marilyn
Monroe and the sudden fall of the studios. This period was certainly a
watershed: nearly all the stars reached their maturity as actors and their
peak as celebrities before 1945; the major studios all began long before
1945; the genres (with the exception, perhaps, of the science-fiction and
horror genres) reached their richest forms in this period; in those sixteen
years, the movies averaged 80 million in weekly attendance – a sum
representing over half of the US population; the movies attracted 83 cents
of every US dollar spent on recreation; the cream of European film talent
had been forced to seek sanctuary in Southern California; and television
had not yet emerged as a serious threat to cinema's popularity (see
Steinberg 1982: 44–9). Norma Jeane Baker thus grew up in a Los Angeles
full of film people, film products, film problems. A generation of budding
Grables and Gables shared the same past, or lack of past, somewhere
within the even heat of Hollywood.

America emerged from the war stronger than before, having escaped
fighting on its home ground and now in the midst of an economic boom,
and in possession of the most powerful weapon on earth, the atom bomb.
Talk of a new regnum, an 'American Century', was not far-fetched. In
1946, domestic films grossed $1.7 billion, the most profitable year in the
industry's history. It was the year that Norma Jeane Baker had her first

screen test at Fox, and experienced the insecurity borne from an awareness of the extraordinary competitiveness amongst hopeful starlets. Hollywood's international dominance made its movies seem mere applications of some given definition of the cinema itself. With the congressional anti-communist witchhunts (1947), the anti-trust ruling (1948), and the loss of overseas markets due to European import tariffs and freezes on the removal of revenues (1947–50), Hollywood became ever more conservative, minimizing risk by keeping to well-tried formulae and well-known stars. The newly named Marilyn Monroe faced a long struggle to even secure a contract with a major studio, let alone start to realize her hopes in the 'Dream Factory'.

Several factors precipitated the crisis which forced Hollywood to search for new stars and new symbols: notably, increasing competition from television and a decreasing stock of plausible plots. In 1946, there were only 8,000 American households with television sets; by 1956, there were 35 million. In 1951, televisions outsold radio sets for the first time. Post-war Hollywood, though not immediately thrown into economic crisis, gradually came to reconsider its role. America had won the war, but in doing so it had lost its splendid isolationism, the freedom to lead a remote existence interrupted only occasionally by threats which could always be quickly defused. America was now forced to admit that its economic reach was worldwide: it was *responsible*. With fewer budget movies with specifically 'American' subjects, there were fewer opportunities to show off promising players to the international public. Only the most magnetic or persistent could hope to achieve anything approximating stardom. By 1953, significantly Monroe's most successful year, 45 million people had stopped going regularly to the movies. Aside from television, this fall was exacerbated by the fragmentation of the mass audience into specialized sections. John Houseman said (Houston, 1963; 169), 'Most of us face this harassing dilemma that we are working in a mass medium that has lost its mass audience.'

It was with this crisis that Marilyn Monroe emerged to be the saviour of the 'goldmining camp and ivory ghetto'. The faded voice whispers, the doll face disowns the well-developed body, almost psychotically impervious to the sexuality below it. The wide eyes blink and squint, in touching, transparent, attempts at cunning. The walk, seeming like the asking of a question. She was the 'bit-part' moved centre-fold, centre-stage, centre-screen: the standard 'dumb blonde' converted into a charming national heroine. Hollywood filled the fifties with a series of challenges to television's success: colour and crowds and curious visual effects – chariot races, the burning of Rome, the dividing of the Red Sea, the crucifixion, and an ice-cream at the interval. The 'sex symbol' served up to entice the all-American male was Marilyn Monroe, a woman who wanted to be loved by you (just you) and nobody else but you, a woman whose profile was projected above you on enormous screens and whose voice whispered sweet nothings in stereophonic sound.

If a woman wanted to succeed in the movie capital, she was obliged to catch the eye of the man with power. Monroe began at the lowest level in Hollywood, scraping a living from part-time jobs and, it seems, part-time affairs (see *Empire News* 9 May 1954; Pepitone and Stadiem 1979). She enrolled in numerous acting, dancing and singing classes, and made good use of the admirers, advisers, patrons and protectors – the most significant of them being the agent Johnny Hyde. She was one amongst thousands of starlets at the Fox studios, many of them dreaming of one day replacing the current 'glamour queen', Betty Grable – a symbol of wholesome sex, combining physical attractiveness with girl-next-door appeal. If one emerging trend can be identified as a factor in Grable's decline, it was the changing social climate in the country: a new frankness about sex was surfacing. A new kind of sex symbol was needed for the new era, giving the film industry the kind of boost it desperately needed.

Fox were in no mood to simply award a good role to an untried starlet. Monroe was placed on the production-line for potential screen stars: she was photographed, sent on promotional tours, and given occasional small parts. Her first screen credit was for a 1947 Betty Grable musical, *The Shocking Miss Pilgrim* (which was only discovered recently). Monroe plays a switchboard operator, with very little to do or say. The movie has rarely been seen, and Monroe's biographers do not seem aware of its existence. Her first remembered role, in *Scudda Hoo! Scudda Hay!* (1948) offered her the 'opportunity' to steal the show with her solitary line: 'Hallo'. The line was cut from the film, although it was reinserted after her death in certain European prints. A further period of demoralizing unemployment was ended by a featured role in a Columbia musical, *Ladies of the Chorus* (1948), in which Monroe made an encouraging performance which revolved around her singing, 'Every Baby Needs a Da Da Daddy'. Tibor Krekes, in *Motion Picture Herald,* gave Monroe her first mention: 'She is pretty and, with her pleasing voice and style, she shows promise.'

By her own account, Monroe (1974) was in the local Schwab's drugstore when she heard that a 'sexy girl' was needed for a walk-on part in a Marx Brothers film called *Love Happy* (1949). Monroe went to the set, met first the director, then Groucho and Harpo Marx. 'Can you walk?' asked Groucho. 'This role calls for a young lady who can walk by me in such a manner as to arouse my elderly libido and cause smoke to issue from my ears.' Monroe walked, and later recalled Groucho saying

' "You have the prettiest ass in the business." I'm sure he meant it in the nicest way.' Groucho announced to the press that Monroe was 'Mae West, Theda Bara, and Bo-Peep all rolled into one', and the scene was shot the following morning. The Monroe character walks over to Groucho and begs for help: 'Some men are following me!' The appearance caught the eye of the critics and, importantly, the attention of the studio.

Love Happy was a fortunate film for Monroe. She was given a role which filled in the space between the narrative action, thus leaving her presence as the sole object for scrutiny. Her distinctive walk was a natural one (see Chapter 2: 'I just walk. I've never wiggled deliberately in my life, but all my life I've had trouble with people who say that I do wiggle deliberately'), although in this context it assumed almost a starring role. It is not her voice that communicates, but her whole body: eyes, smile, hair, gestures, curved hips, swaying walk. The body is savoured, received and displays its own discrete story. Her clothes were carefully chosen to highlight the curves of the body, and their colour (light) contrasted with the men's black suits. In movies, clothes *are* the body, as the expression is the face: in the case of Monroe, this fact assumes a particular pertinence. In movies clothes conceal; hence they conceal something distinct from them; the something is therefore to be uncovered. A woman in a movie is *dressed* (as she is, in reality), hence potentially undressed. One's eyes are invited to move about the body. The ontological conditions of the motion picture reveal it as inherently pornographic (though not of course inveterately so). Lovers close their eyes the moment before they kiss, flinching from the distraction of a mutual close-up. The camera is a less romantic suitor, seeking to see everything, all of it, all the way through. This ceaseless ingestion of images is only frustrated by extra-cinematic concerns. The instances seem endless in which a scene stops the instant the lips touch, the zipper slides down the back of a dress, or in which the woman steps behind a shower *exactly* as her robe falls, or in which an item of clothing drops tantalizingly into the view of a suddenly paralysed camera. These are not sudden enticements or pornographic punctuations; they are satisfactions, however partial, of an inescapable desire. All the Hays Offices and 'moral majorities' were powerless to prevent that; they could only insist upon the interruptions (see Cavell 1979).

The distribution of emphasis between a man and a woman in a movie is like its distribution in a *pas de deux*. The most durable male star cannot survive a cipher as his object (witness Cary Grant with Monroe in *Monkey Business*), and if one is for some reason repelled by a given woman actor, no man can rescue the movie. The leading woman has the eyes of the camera, and the audience is obliged to see eye to eye with the camera or else refuse to see it through. In Monroe's early, 'pin-up' period, her screen appearance is most clearly dictated by conventional Hollywood conceptions of woman's 'proper', peripheral role – short on substance but strong on spectacle (as anyone misfortunate enough to view Anne Baxter's role in *The Ten Commandments* – 'Moses, Moses, you splendid, stubborn, adorable fool' – will surely confirm). The early appearances by Monroe have her cast, archetypally, as the 'girl' defined solely by age, gender and attractiveness. In *Scudda Hoo! Scudda Hay!* and *Love Happy* she does not even possess a name, and in *Dangerous Years* (1948), *The Fireball* (1950) and *Right Cross* (1950) her character has no 'biography' beyond being 'the blonde'. Her characters' jobs in this period include chorus girl (*Ladies of the Chorus, Ticket to Tomahawk*, 1950), struggling 'starlet' (*All About Eve*) or (often ill-qualified) secretary (*Hometown Story, As Young As You Feel, Monkey Business*). Reviewers describe her at this time as 'a beautiful blonde' (*Asphalt Jungle*), as 'curvy Marilyn Monroe' (*As Young As You Feel*), and as having 'a shapely chassis' and being 'a beautiful shapely blonde' (*Let's Make it Legal*). It is evident that the critics, who seem to have a morbid fascination with 'shapeliness', could only perceive the young Monroe in her function as embodiment of sex appeal: 'Marilyn Monroe is tossed in to cause jealousy between the landlords' (*Variety*). 'Leatrice Joy . . . gives mature warmth to the proceedings. Marilyn Monroe has that other quality' (*Film Daily*). The gentlemen of the press, visiting the lot of *Clash by Night* (1952), were overheard by the star, Barbara Stanwyck, as saying: 'we don't want to speak with her [i.e. Stanwyck]. We know everything about her. We want to talk with the girl with the big tits.'

Fox chief Darryl Zanuck ordered Monroe to be written into any picture which could 'use' a sexy blonde. Each minor role brought an extraordinary number of costume changes, and whenever a convenient lull in the plot occurred she was given some extraneous primping in mirrors to extend her role. Zanuck was now interested, and this interest did not allow Monroe's personal identity to freely develop; as Milton Sperling notes,

If you were a Warner employee, or a Fox employee, or a Metro employee, that was your home, your country. You voted en block for your company's films at the Academy Awards. . . You had T-shirts with your studio's name on them. It was just like being a subject, and a patriotic subject at that. People who lived and worked beyond the studio walls just didn't belong, and you were prepared to fight them off, like the Phillistines. (In McClelland 1985)

Zanuck's studio saw Monroe as a potential 'blonde bombshell', a potent and profitable new version of a standard type: the dumb blonde. The hair of the platinum blonde was a super-textured, 'more blonde than blonde'. It was also, to many, an image of the white race that excludes all other races. Fox had an established policy of promoting blonde women stars (Faye, Grable, Havers), and blondeness was equally commonplace in pin-up photography. Allied with blondeness was innocence, a childlike quality which Helen Kane – the 'boop-boop-a-doop girl' – affected in the twenties, eventually inspiring the cartoon Betty Boop ('Just a perfect little she'). The 'child–woman' image, an intrinsic part of the dumb-blonde type, was used as a source of light comedy (e.g. Betty Boop); the image gradually moved from showing female attractiveness as childlike to showing women's own sexual feelings as childlike or infantile (e.g. Clara Bow). The dumb blonde type also drew on the 'dizzy comedienne' role, a form of vaudevillian humour popularized by Gracie Allen and ideal for giving witticisms to women whilst denying their creativity ('Ha! That crazy dame!'). Combined, these traditions coalesced in the type which revealed, and ultimately restrained, the qualities attributed to Marilyn Monroe – the bright, bubbly, blonde sex symbol, flirtatious and fun and fetchingly free.

Monroe had already had half a dozen small parts when she walked through *All About Eve* (1950), leaving behind one of the first authentic 'Monroisms'. As Miss Caswell, a starlet, she arrives at a party on the arm of stage critic Addison DeWitt (George Sanders) who suavely introduces her to Margo Channing (Bette Davis), 'You know Miss Caswell, of course?' 'No!', Margo mercilessly snaps back. 'That', smiles Miss Caswell guilelessly, 'is because we've never met.' The public had begun to notice her, the press became used to the regular supply of studio stories on Monroe, and her roles became bigger and more significant. As her successes became more frequent, Hollywood found itself confronted with a classic dilemma: as a capitalist enterprise, it was obliged to pursue profits, and Monroe's sex appeal was proving profitable – yet this sex appeal conflicted with prevalent beliefs and values amongst important pressure groups. The studio, as producer of commodities and producer of ideology, tried to solve the problem by attaching Monroe's image to certain, ideologically acceptable images of female sexuality: coping with the threat of sex by displacing it (glamour), excusing it (the 'dumb blonde' syndrome), or punishing it (the *femme fatale* image, the suffering motif). These strategies only worked ambivalently, monitored by a manufacturer in the ambivalent position of purveyor of sexuality and protector of morality.

Whereas the publicity worked to maintain this uneasy balance ('Marilyn visits the children's home'), the promotion of her movie image worked against the balance: the starlet was elevated to the level of 'goddess'. There was, it seems, a particular constellation of qualities possessed by Monroe which prevented her being moulded into anonymity, in the way that other pin-ups such as Lana Turner were, or crushed into obscurity, like Frances Farmer. Whilst publicity sought to strengthen Monroe's stereotypical image, her promotion and her performance subverted this stereotype. In her early period the studio was caught in several minds as to an appropriate handling of her sexuality: it made her 'bad' by putting her in *films noir;* it later made her 'respectable' by associating her with prestige. *Don't Bother to Knock* (1952) and *Niagara* (1953): *films noir* fables. Socially the war had brought women into the labour market and disrupted old sex roles; culturally this disruption was treated as a sexual challenge. 'White women' suddenly seemed 'Red': the old order was disturbed. Sexuality offered men short-term pleasure but long-term pain; women were desirable but destructive, dangerous to men. Women are punished for their imputed evil or illness: the presentation of this punishment will usually gratify the sadism of the male viewer, directly and through his surrogate, the male protagonist (see Mulvey 1975). Monroe was also given 'prestige' musicals with successful Broadway pasts, and she was given an enhanced respectability by her association with 'name' directors – Huston, Lang, Mankiewicz, and later with Wilder, Preminger, Logan, Olivier and Cukor – and front-line stars – Cary Grant, Lauren Bacall, Charles Laughton, Bette Davis, Clark Gable.

A significant example of the 'victim' plot is *Don't Bother to Knock*. Monroe's own fragility is here presented, for the first and really the only time in her career, as psychotic. She plays a baby-sitter (named, with Dickensian symbolism, 'Nell') whose lover was killed in a plane crash. She has never recovered from the trauma of his sudden death. 'Everybody tries to come between Phillip and me,' she whispers, eerily. She tries to kill the child she is minding, accusing the girl of coming between her and her men. 'Phillip is dead. Do you know it?' asks Richard Widmark, as a man she meets in the hotel where she is working. Traumatized, self-abnegating to a pathological degree, unreachable, the woman lives in a closed fantasy world of second-hand clothes and sordid dreams: 'I'll be any way you want me to be. From the beginning, I knew you were the very best. Don't leave. I was in a hotel room once, the night before he flew away, for the last time.' As this lost child-woman, hopelessly alienated, sensitive, strange, inarticulate, Monroe is interestingly cast, revealing dark aspects of her own tortured personality. Anne Bancroft, playing opposite her in one scene, witnessed Monroe playing someone experiencing a breakdown: 'It was so real. . . I really reacted to her. She moved me so that tears came into my eyes.' As she is taken away to an asylum at the end, her last words are a fragment: 'People who love each other. . .' Her collapse to madness seems absolute, terrifying in its finality, and it shocked the critics. Comments ranged from the patronizing (Monroe 'has all the right measurements but not all her marbles', *New York Herald Tribune*) to the dismissive ('unfortunately, all the equipment that Miss Monroe has to handle the job are a childishly blank expression and a provokingly feeble, hollow voice', *New York Times*). Although the reviews discouraged audiences from seeing the film, it now appears as an interesting early instance of Monroe's personal inflexion of a role; her acting is erratic as she struggles to play a 'vacant woman' without seeming wooden.

Monroe's studio tried one last time to 'punish' her sexuality, and did so with rather more success. *Niagara,* directed by Henry Hathaway, cost $1,250,000 and enjoyed a return of $6,000,000. Joseph Cotten plays a man who 'went wrong' in the war. In the haunting first scene, he is 'called' by the falls, to commune with their titanic force. He is linked to them throughout the film, as if they symbolize the forces churning within him. His wife, played by Monroe at her most brazen, is a typical *noir* figure – a woman whose sexuality attracts men only to destroy them like a fly devoured by a spider. He suspects her of plotting to kill him, and he is tormented by her indifference. Dressed in a figure-hugging red silk dress, she wanders out and requests that a party plays 'Kiss' – 'the *only* song', she says. Two young newly-weds, not so much clean-cut as ready-scrubbed, watch in astonishment as the *femme fatale* basks in the aura of her own sexuality: 'Kiss me . . . take me in your arms and make my life perfection. . . *Perfection.*' Her husband rushes out from their room and smashes the record, badly cutting his hands. Hathaway's obvious amusement at this scenario rubs against the film's dramatic development, ironically serving to heighten the shock impact when the man kills his adulterous wife and chooses his own destruction by going over the edge of the falls. A character to whom something awful – something irreversible – has happened, he is sick beyond cure and perhaps beyond explanation. Monroe's 'victim' was eminently impressive, made particularly attractive by attentive camera direction, rich production values (lush colour and eye-catching costumes), and a song motif. *Niagara* also firmly established 'the Walk', with a 70-foot shot of Monroe undulating away from the camera in uncomfortable high heels across cobble-stones. The walking woman, moving away from us, moving rapidly, rhythmically, revealingly, making her figure do precisely what seems most significant, yet retaining a curious aura of innocence. Margot Fonteyn (1976: 184) met Monroe around this time and noted:

> She was astoundingly beautiful. . . What fascinated me most was her
> evident inability to remain motionless. Whereas people normally move

their arms and head in conversation, these gestures, in Marilyn
Monroe were reflected throughout her body, producing a delicately
undulating effect like the movement of an almost calm sea. It seemed
clear to me that it was something of which she was not conscious; it
was as natural as breathing, and in no way an affected "wriggle", as
some writers have suggested.

Monroe became, in 1953, a Hollywood star. She was receiving more
than five thousand fan letters a week, and was now to occupy a luxurious
dressing room formerly inhabited by Marlene Dietrich. The studio began
calling her '*Miss* Monroe'. Her marriage to 'Jolting Joe' DiMaggio only
reinforced her popularity. She was finally awarded a 'tribute' whose
tastelessness was only matched by its loudness: a song called 'Marilyn':

No gal, I believe,
Beginning with Eve,
Could weave a fascination like my MA-RI-LYN
I planned everything,
The church and the ring,
The one I haven't told it yet is MA-RI-LYN.
She hasn't said yes,
I have to confess,
I haven't kissed or even met my MA-RI-LYN
But if luck is with me,
She'll be my bride for evermore.
(Starlight Songs Inc., 1952)

Monroe later reflected,

I really got the idea I must be a star, or *something,* from the
newspapermen . . . who would interview me and they would be warm
and friendly . . . they'd say, "You know, you're the only star", and I'd
say, "star?" and they'd look at me as if I were nuts. I think they, in
their own kind of way, made me realize I was famous. (In Meryman
1962)

By the end of 1953, according to the trade press, Monroe had made
more money for her studio than any other female star in Hollywood.
After success in *How to Marry a Millionaire* (in which, symbolically, she
plays alongside her predecessor, Betty Grable) and *Gentlemen Prefer
Blondes,* she won an award as 'the year's most popular actress'.
Internationally, Monroe was making an astonishing impact; according to
the *Daily Herald,* 'In Turkey a young man went so daft while watching
Marilyn wiggle through "How to Marry a Millionaire" that he slashed

his wrists.' The news reappeared in *Film Fun*.

Gentlemen Prefer Blondes is one of the films most often invoked when considering Monroe's image, and it thus demands serious attention. The film, adapted (via Broadway) from the Anita Loos novel, was purchased for Betty Grable. Monroe won the part, not only because her star was in the ascendency, but also due to her fee being $15,000 compared with Grable's $150,000. Monroe was an obvious choice to play Lorelei Lee, the dumb blonde who knew that diamonds are a girl's best friend. Loos allowed the character to be redrawn as a 'White Marilyn': the original Lorelei of myth carried 'Red' connotations, as she combed her long blonde hair and sang, luring sailors to wreck their ships on treacherous rocks. 'These rocks don't lose their shape,' sings Monroe's Lorelei, but she is depicted as too gentle a figure to pass from flirtation to threat. She fully understands how to use her sexuality to trap rich men ('a man being rich is like a girl being pretty') and is motivated above all by cupidity. Her dialogue as written is self-aware and witty, signalling amusement at her actions even when she is playing the *fausse-naive*. The weight of the Monroe image, in contrast to Lorelei, is on innocence. She is certainly aware of her sexuality, but she is guiltless about it and it is, moreover, presented primarily in terms of narcissism – sexuality for herself rather than for men. There is thus a disjunction between Monroe-as-image and Lorelei-as-character. They only touch at three points: the extraordinary impact of their physicality, a certain childish manner, and a habit of uttering witticisms. The general disjunction thus serves to confuse the narrative (are we watching Lorelei or Marilyn?) whilst highlighting the star status of 'Marilyn'. Indeed, when Jane Russell's character is obliged to impersonate Lorelei, she simply presents all the traits we the audience associate with Marilyn Monroe.

The film, a musical-comedy, is a self-consciously 'quality' production, meaning expensive sets, vivid colours, extravagant costumes and a famous cast. The musical numbers are manipulated into contributing to the characterizations and narrative development. 'A little girl from little rock' is given several inflections: a marching tune (the two women join the male athletes on board ship), a swing number (the 'girls' celebrate togetherness) and a raucous finale (the story has resolved itself). The moral is, like our two heroines, that the basic song remains the same despite the pressures of context. Monroe and Russell are thus attuned to each other's needs: they are *strong* women, they clearly care for each other. Monroe's character is awarded the key dance number, 'Diamonds are a Girl's Best Friend'. She sensed the significance of the song for her role, and insisted upon retake after retake in order to 'be better'. The first take was in fact the one chosen, and it is a memorable celebration of Monroe at her peak.

Although Jane Russell and Monroe were co-stars, Monroe is the sole star for the camera. She is given the most striking costumes (usually with plunging necklines and bright colours – contrasting with Russell's dark, 'mannish' clothes) and is generally centre-frame. Background music tends to become active when Lorelei enters the frame, and the plot weights audience interest on Lorelei's prospects. It is an impressive result from a collaboration between a talented, self-aware star and a production team keen to promote that star. The afterglow from the film's success helped to protect Monroe's reputation after two uninteresting pictures: *There's No Business Like Showbusiness* (1954), a tiresome 'revue' featuring a titillating version of 'Heat Wave', and *River of No Return* (1954), a crude exploitation of sexual attractions. It was becoming evident that not only was Monroe tiring of her 'dumb blonde' roles, she was also resentful of studio publicity which depicted her as a fairly frivolous, fairly foolish young woman. She told Maurice Zolotow (1961: 77):

> I'm nobody's slave and never have been. Nobody hypnotizes me to do
> this or that... Do you ever see on the screen, "this picture was
> directed by an ignorant director with no taste?" No, the public always
> blames the star. *Me.* I had directors so stupid all they can do is repeat
> the lines of the script to me like they're reading a timetable. So I didn't
> get help from them. I had to find it elsewhere.

Monroe sought help from drama coaches Natasha Lytess, Michael Chekhov and, later, Paula and Lee Strasberg. Her attempt to improve and develop as an actor was greeted with heavy sarcasm by the press, and by her own studio. Increasingly she found herself imprisoned by her fame: attendance at evening classes was made impossible after her identity was made public, and her presence amongst the intellectual elite was never entirely welcome. Yet Monroe enjoyed many aspects of her star status, and it was certainly a formidable structure to struggle against. She was a hugely successful stereotype, because she was *more* than a stereotype: this is the paradox both she and her studio puzzled over. In the stereotyping process, Hollywood publicists worked with studio policy-makers to assure that their efforts were consistent with the screen image. The publicity build-up started months before Monroe made her next movie, *The Seven Year Itch* (1955). The publicity (concerning the film, Monroe's involvement, her co-stars, her marriage, her social appearances) found an outlet in syndicated Hollywood gossip columns and fan magazines. When her role was finalized, the studio assigned a 'unit man' to 'plant' items related to her and her role in the press; television appearances were discussed, interviews were arranged. New York publicity offices of the studio took over the campaign when the distribution-exhibition stage started; national advertising and merchandise tie-ins exploited the film's promise. The publicists worked on the idea of Monroe as the symbolic object of illicit male sexual desire. They had a gigantic blow-up of her *Seven Year Itch* pose positioned over a street. They concentrated on her breathy voice, her 'horizontal walk', her distinctive white dress, her half-closed eyes and half-opened mouth. She was consistently typecast as to arouse certain expectations from her public, making it hard to successfully assume a new change of status and still maintain her popularity with the public which elevated her to a position that enabled such changes. This dilemma disturbed Monroe throughout the making of *The Seven Year Itch;* it was her final studio-monitored film.

The Seven Year Itch, directed by Billy Wilder and co-written with George Axelrod, provided a euphemism for (predominantly male) sexual infidelity and emotional insecurity. Ironically, no such infidelity occurs in the film, and the characters are not so much insecure as immature. The theme explored concerns the enslavement of the modern American male to media-manufactured fantasy women. There were serious problems: severe censorship restrictions manoeuvred the writers towards comedy as a surrogate to eroticism; Monroe's star status entailed a lesser-known male-lead (Tom Ewell), reducing the potential sexual and dramatic tension. 'This has never happened to me before,' splutters Sherman (Ewell) after a clumsy pass on the piano stool, to which 'The Girl' replies, 'Oh, it happens to me all the time!' Her innocence exacerbates his crisis, and he wanders off into adulterous fantasies lifted from the movies. While Wilder's dramatic instincts push in one direction, the censorship codes pull in another; when the tension becomes acute, a knock on the door or a spilt drink saves the play. The male character suffers ceaseless psychological and cinematic abuse for the carnal contemplation of Monroe, and the audience is vicariously judged and chastized for its voyeuristic participation.

Monroe is made to appear childlike – evidently by her girlish joy as the wind from a subway ventilation shaft causes her dress to billow around her waist – and sexually unattainable – 'In this heat I always keep my undies in the icebox.' The film's ideological obsession with rendering Monroe innocent and frigid, together with the blatant phallic symbolism with the subway train which raises her white dress, makes Monroe into a fantasy figure beyond the reach of men. Indubitably, inevitably, *nothing is going to happen.* Wilder later acknowledged his difficulties: 'Unless the husband left alone in New York . . . has an affair with the girl there's nothing. But you couldn't do that in those days, so I was just straightjacketed. It just didn't come off one bit and there's nothing I can say about it except I wish I hadn't made it.'

Despite the dramatic weaknesses, *The Seven Year Itch* is a fascinating instance of Monroe-as-spectacle. She is struggling inbetween 'The Girl' as a type and as a novelistic character: one is typical (the studio's choice), one is unique (her choice). 'Isn't she a livin' doll!' The 'typical' blonde has few traits, is immediately indentifiable, has a pre-given personality, does not develop with the narrative, and indicates 'society'. The 'novelistic character' has a multiplicity of traits, her identity is gradually revealed and develops with the narrative, and indicates 'the individual'. In the film, 'The Girl' seems to still stand out as a type *and* to offer some kind of comment on it. Monroe brings an irony to the part which another woman, with less public 'history' and a less prominent image, could not match: 'That's the wonderful part about being with a married man. No matter what happens he can't possibly ask you to marry him, because he's married already, right?'

The single, short scene which has burned into the public memory of Monroe is the 'subway' scene with the billowing white dress. Ironically, the famous shooting of the stunt, at 52nd Street and Lexington Avenue in New York, failed to provide the desired effect – the printed version was shot in the studio, first take. Monroe stands over the grating and sways with pleasure as the cool breeze blows around her legs; her male companion looks on admiringly in the background. As with any image, the meanings do not inhere solely in the gestures, *mise-en-scène,* colour of the image *per se,* but in what is known about its production, its surrounding publicity, what it has been taken to mean subsequently. For example, the childhood anecdote given by Monroe (1974: 16), whether it really originated with her or with her ghostwriter Ben Hecht, certainly seems significant when one watches the rising skirt: 'This wish for attention had something to do, I think, with my trouble in church on Sundays. No sooner was I in the pew with the organ playing and everybody singing a hymn than the impulse would come to me to take off all my clothes. I wanted desperately to stand up naked for God and everyone else to see.' Contemporary audiences had already read about the scene, and images were incorporated in related advertisements; we now know of the difficulty the scene caused and the anger it caused Monroe's husband. Generally, that image of Monroe emphasizes the sense of narcissism, the curious mixture of delight in one's own physicality and one's awareness of sharing it with the audience. In the context of the film's plot, this is the metaphorical 'exposure' the male character (and, in fact, male viewers) have been seeking; its displaced quality – the lack of a man to affect the action – is due to the censorship restrictions, and the fact that no respectable married man who entered into an adulterous affair with

Hollywood's ultimate sex-object would be tolerated by contemporary audiences.

Charles Schnee summed up a common feeling when he wrote:

> We've been soft-pedaling on sex here in Hollywood, and now who've
> we got besides Marilyn Monroe who as a single feminine star-
> attraction can pack the house? This Hollywood habit of frowning on
> out-and-out sex appeal in an actress is costing the industry some
> much-needed star power for its pictures. . . In recent years we have
> gone almost puritanical, and now, outside Miss Monroe, we haven't a
> single new actress who on sheer strength of feminine appeal can
> attract, unfailingly, huge throngs at the theatres. (In McClelland 1985:
> 43)

Hollywood was now in serious trouble. Several studios went under, and the major studios financed independent producers who in turn were releasing their films through the financing studio. It was evident that the independent producer was becoming the most important person in films. Monroe set up her own company and left Hollywood for New York, hoping to gain control over her movie vehicles and to improve on her performances. She said of her studio: 'They're screwing me with that salary and I don't like to be screwed.' Asked by the press as to why she began learning acting she replied, 'Seeing my own pictures'. She went on to say:

> I want to develop. I just want to grow. . . I want to grow in stature, to
> be a real actress. In New York I learned to make friends. Before, I
> never had any friends, only conquests. I didn't have the time to find
> real friends. I was always being looked at, had no chance to look. I am
> perfectly serious about wanting to act seriously. (In McClelland 1985:
> 137)

Her next performance, directed by Joshua Logan, was *Bus Stop* (1956). She had learned from the Actors Studio the 'Method' of striving for 'authenticity' in characterization. As she familiarized herself with Stanislavsky's 'associative' thinking, her speech assumed a more poetic quality, regularly employing similes and metaphors in order to express her feelings. She also took more interest in dress as an aspect of communication. For her role as Cherie, a 'chantoose', Monroe sought out old costumes, made holes in her stockings, and used a chalk-white make-up which photographed as a pasty complexion. She sang out of key during her dance number, remembering that Cherie was a rather poor performer. Despite some anxiety amongst studio bosses, Monroe produced a comic performance of flair and charm and pathos that transcends the stage-bound screenplay. The fact that she did not recieve so much as an Oscar nomination surprised many critics, and further reinforced Monroe's feelings of rejection by the cultural elite.

Monroe felt the bars of the prison-house closing in upon her; she told Weatherby (1976: 150):

> When I was a kid, the world often seemed a pretty grim place. I loved to escape through games and make-believe. You can do that even better as an actress, but sometimes it seems you escape altogether and people never let you come back. You're trapped in your fame. Maybe I'll never get out of it now until it's over.

Now married to playwright Arthur Miller, Monroe found herself under pressure for political reasons. Miller was under investigation by the House of Un-American Activities Committee (HUAC), and movie bosses told Monroe that she would 'never work again'. The *Daily Express* (27 June 1956) reported:

> One way for an actress to get herself disliked in America at this moment is for her to have any connection with Communism – even some remote "guilt by association" is enough. Women's clubs and ex-service organisations would boycott her films. Studios suddenly discover they have no parts for her... It could happen to Marilyn Monroe.

Undeterred, Monroe bought the rights to Terence Rattigan's stageplay of Cowardly camp, *The Sleeping Prince*. Her appearance alongside Sir Laurence Olivier in the film version, *The Prince and the Showgirl* (1957), her only British film, seemed the culmination of her cultural ambitions. However, Olivier's direction was diametrically opposed to Monroe's 'Method' technique, and her increasing periods of illness only exacerbated his frustration. One morning he walked on the set and instructed his star: 'Okay Marilyn, be sexy.' A tense Monroe failed to detect the (presumed) humour in the request; she was aware of Olivier's belief that Vivien Leigh, his wife, should have been given the part. He tried to calm her by advising her to count slowly to herself. She remained tense; he shouted 'She can't even count!' Despite Olivier, she acted. Most strikingly, her performance alone in the room after her night of (imagined?) romance is an extraordinary demonstration of suggestiveness and subtlety. She is shown radiating her novel experience of her body, revealed in a restrained series of cuts: the progress of her first habitual exploration of one person, the way her body has become marked and scented and charted with the knowledge of his, containing her within a fresh skin of mood and gesture and odour and boredom and bristle which she feels over her neck as she reclines on the chaise lounge, in her mouth as she takes her tongue over her lips, in her shoulders and thighs and legs as she rolls over and relaxes, in her hands and nose and teeth as she sips champagne from a chilled glass. It is a quite awesome projection of personality, quite refuting Olivier's scepticism. The overall impression one takes, watching this conscious spectacle, is a reflection of Dame Sybil Thorndyke's comment on Monroe: 'We need her desperately. . . She's the only one of us who really knows how to act in front of a camera.'

Oliver's temperament was clearly ill-suited to the Method dictum, 'Use your past, use your pain.' Surely, he insisted, it is all just acting, doing a job. His character, the Grand Duke, seems fussy, brittle, artificial – certainly not impressive enough to satisfy the demands of the plot. Monroe's character, on the contrary, seems delicate and finely drawn, the work of someone well aware of how to exploit the specific qualities of the movie camera. Monroe seems to hide behind nothing, revealing herself both physically and emotionally: she wears a revealing white dress and speaks the way she always speaks. Movie acting was becoming more instinctive for her, yet the professional projection of personalities, the incessant role playing, was becoming more disorientating.

Monroe was now dependent upon painkilling drugs and frequently prone to bouts of extreme depression. Simone Signoret, who knew Monroe during the final few years, believed that her confidence had gradually been eroded. Signoret writes (1979: 322–3) that Monroe began to dislike her work:

> She didn't like it very much because there had been a whole succession of people in her life who had taken pains to explain to her that she was anything but an actress. They had made her believe that without them she was incapable of saying "it's going to rain" and have it ring true in any way. . . They had thought that the starlet Marilyn was cute. They detested her for becoming Monroe.

It was debatable whether the demands of movie acting were helping or harming Monroe's self-belief. She said,

> A struggle with shyness is in every actor more than anyone can imagine. There is a censor inside us that says to what degree we let go, like a child playing. I guess people think we just go out there, and you know, that's all we do – just do it. But it's a real struggle. I'm one of the world's most self-conscious people. I really have to struggle. (In Taylor 1983: 70)

Billy Wilder's *Some Like It Hot* (1959) enabled Monroe to confirm her eminence as a gifted comedienne; it also exacerbated her emotional problems, adversely affected her marriage, and caused her to suffer a miscarriage during the filming. Her illness made her lateness into a major production hazard. Commonly her behaviour was interpreted as 'Norma Jeane's revenge' on authority; it was nothing of the kind. Nembutal sleeping pills would often leave her so incapacitated she would be made-up whilst she was lying flat in bed. She was known to pass out or vomit or become too dizzy to act. Make-up man Allan Snyder says that there was never a time when Monroe did not feel 'terror, pure terror' about acting. His colleague, wardrobe assistant Marjorie Plecher, says, 'She was so scared about looking right, acting right, that she was physically unable to leave the trailer. It was the ultimate stage fright. She had a great talent, but she never felt sure of herself, never could believe in herself' (in Summers 1985: 81).

Monroe struggled to perfect one of the most notable film comedy performances, achieved by multiple retakes during which she became better and better, but which co-star Tony Curtis believed had destroyed his 'spontaneous' style. 'Before each take,' wrote Lloyd Shearer, 'Marilyn would close her eyes and enter a deep trance. She would pull down on her creeping-up bathing suit, style 1927, then suddenly start to flail her hands violently, up and down, as if she were desperately intent upon separating her hands from her wrists.' (*Parade* 7 December 1958.) Curtis later attacked Monroe for her 'vicious arrogance', and said that kissing her in one long scene had been like 'kissing Hitler'. It was a strong and very sad statement coming from a Jew, and has not been tempered by time. Jack Lemmon, Monroe's other co-star, gave me a rather more sensitive account:

> You're right to stress her personal problems. She was a sweet lady who was clearly going through some kind of hell on earth. I didn't know all the reasons, but I saw she was suffering – suffering and still producing that magic on film. It was a courageous performance, really courageous. Most actors only occasionally use all their talent, but Marilyn was using hers *constantly*, giving everything she had, till it hurt, struggling to be better. Sure, it was infuriating for us, at times, but I was really fascinated to watch her work.

The director, Billy Wilder, seems to appreciate the specificity of Monroe's predicament: 'I have an aunt in Vienna, also an actress. Her name, I think, is Mildred Lachenfarber. She always comes to the set on time. She knows her lines perfectly. She never gives anyone the slightest trouble. At the box office she is worth fourteen cents. Do you get my point?'

The basic plot of *Some Like It Hot* concerns two struggling jazz musicians who inadvertently witness a gangland massacre, perpetrated by Spats Columbo and his cohorts, in the Chicago of the twenties. It is the St Valentine's Day Massacre, when men slaughtered men on the day celebrating love. In order to escape the vengeful wrath of Columbo's gang, they disguise themselves as members of an all-girl jazz band, Sweet Sue's Society Syncopaters, who are on their way to Florida. Disguise is impelled by physical fear, the two men literally petrified into women. Complications ensue when Joe (Tony Curtis) becomes attracted to the lead singer of the band, 'Sugar' Kane (Monroe), and has to devise a way of shedding his female disguise; when Jerry (Jack Lemmon) is relentlessly pursued by a smitten millionaire, Osgood Fielding III (Joe E. Brown); and when Columbo's gang reappear in Florida.

All of the characters are seen to play *against* some rival identity. Curtis and Lemmon produce parodies of their screen personalities (Brooklyn Romeo and New York neurotic), and proceed to submerge them beneath their 'female' disguises: 'I tell you', says Jerry, 'it's a whole different sex!' As men, they pose a threat to Spats as witnesses; disguised as women, they become even more threatening to the homophobic mobster. Spats is at his most anxious when his white spats are stained with blood: his obsession with cleanliness and purity suggests that he finds sex and women dirty. His usual order is 'Get 'em, boys!' Jerry at one point laments: 'I'm a boy, I'm a boy. . . I wish I was dead!' Monroe portrays her writers' version of 'Marilyn' (the part was specially written for her), but struggles throughout to present her own personal fiction. What *Some Like it Hot* achieves is an exploration of the transformation in character which Joe and Jerry undergo when experiencing this sexual sea-change, and a satire of the stereotyping of male and female roles, which are taken in logical stages to comically absurd conclusions. This is probably seen at its most poignant in the scene in Osgood's yacht between Joe and Sugar where the erstwhile womanizer, compelled to feign femininity to save his life, and now compelled to feign impotence to further his romance, conducts a seduction scene flat on his back which can only succeed through his playing the passive 'feminine' role himself and being seduced.

Despite his female disguise, Joe nevertheless retains his own identity: he merely moves from Joe to Jo-sephine. However, instead of becoming Geraldine, Jerry walks into the train compartment and announces to the girls: 'Hi! I'm the bass fiddle, just call me Daphne!' The effect on Joe of the role reversal is ultimately to make him more humane; he becomes genuinely moved when Sugar confides in him as a friend. Conversely, the effect on Jerry is to make him enjoy being a girl so much that he becomes engaged to Osgood, can think of no good reason why the wedding should not go ahead, and, to Joe's bemused query of 'Why would a guy want to marry a guy?' can think of the snap reply: 'security'.

Joe labours to conform to the stereotype of the ideal man whom Sugar has described: shy, bespectacled, clever, rich, and inexperienced with women. The image is Wilder's sly reference both to the character of Sugar (her search for a Sugar daddy) and to Monroe herself (married to shy, bespectacled, intellectual Arthur Miller). The audience is constantly invited to tease out the biographical references in Monroe's character. Wilder, desperate to direct Cary Grant but without the opportunity to, gleefully undermines the image of the cinema's most dexterous lover by having his impotent soundalike seduced by the cinema's ultimate sex symbol.

Suddenly discovered by the hoodlums, Joe and Jerry are forced to make a frantic escape. As Daphne/Jerry telephones Osgood to suggest a rapid elopement, Joe is drawn towards the concert stage, seduced and saddened by Sugar's plaintive rendering of 'I'm Through With Love'. Visibly moved, he walks onto the stage, moves towards Sugar, kisses her passionately on the lips and, looking into her tear-stained eyes, whispers: 'None of that Sugar. No guy is worth it.' Thus, Monroe's most memorable screen kiss is given her by a man dressed as a woman. In addition to being one of the most moving moments in Monroe's screen career – the song a poignant testament to her personality both on and off the screen – it is one of the most audacious and yet touching moments in American screen comedy.

The film is so significant for Monroe watchers, for it is the quintessential fiction *on* Monroe. It is evident in every turn and twist in the script. Her image is so highly defined it is impossible to separate 'Sugar' from 'Marilyn'. Sugar's goodness encourages Jerry to empathize with her: 'We wouldn't be caught *dead* with men. Rough hairy beasts! Eight hands!' When Jerry says, 'Hi! I'm Daphne', he effectively says farewell to his masculine self. In Greek mythology, Daphne was a nymph who resisted the atttentions of Apollo by transforming herself into a tree, a transformation lasting for eternity. The analogy is apposite: Jerry's adoption of Daphne is a permanent pose. As the film progresses, he begins to grow more and more like Monroe (at one point, on the beach, Monroe even expresses her admiration for Lemmon's 'figure'). By systematically denying Jerry all traces of masculinity, Wilder prepares us for one of the most celebrated of all screen denouements.

The story concludes with two confessions. Without going into the ramifications of his 'Josephine' disguise, Joe tears off his wig and explains to Sugar that he is a dissolute saxophone player and not a millionaire. Sugar's kiss betokens forgiveness in a way that matches his conciliation and also a submission to her perennial weakness for saxophone players. Daphne has a more difficult problem with Osgood. Daphne explains the obstacles which preclude their betrothal: 'I smoke. . . I can never have children. . .' Osgood is unperturbed – 'Doesn't matter. . . We can adopt some!' 'Oh, you don't understand Osgood!' says Daphne/Jerry, pulling off his wig: 'I'm a man!' Osgood, momentarily put off balance, quickly recovers and continues smiling. 'Well,' he says, 'nobody's perfect.'

Some Like It Hot has Marilyn Monroe give one of her most touching screen performances, as well as her funniest. Sugar Kane is a romantic who is consistently unlucky in love and almost absurdly vulnerable. The audience had followed Monroe's personal highs and lows for several years, and brought them to the movie theatre. Sugar, like Marilyn, is shown as the archetypal 'dumb blonde' who is 'not very bright'; the bighearted victim who 'always gets caught'; the generous creature who always ends up with the 'fuzzy end of the lollipop and the squeezed-out tube of toothpaste.' She says of men: 'You fall for them, you really love them, you think this is going to be the biggest thing since the Graf Zeppelin. The next thing you know, they're borrowing money from you, they're spending it on other dames, and betting on horses. . . Then one morning you wake up, the guy's gone.'

Some Like It Hot is one of those characterizations of 'Marilyn' (*The Misfits* is another, see chapter 6), whose intensity gave the Monroe myth its powerful currency. No other director has more effectively tapped Monroe's sense of fun. She has a blissful vulgarity, an innocent sexiness, a verve and abandon which clearly fascinates Wilder. In one memorable scene, Wilder conspires to show Monroe as, symbolically, she seemed to the audience: as a seductress, whilst neither Joe the 'millionaire' nor the man in the movie theatre can act – both are impotent. Eventually, Wilder relents: Sugar pours herself over the prostrate figure, pressing her flesh against him, softly, strongly, stroking his hair, kissing him first gently, once, twice, several times, then heavily, smothering him, commanding his whole body – from 'a tingling sensation in my toes' to the misting-up of his spectacles. He is being seduced by a movie goddess: 'It's a miracle!' It is not surprising that she succeeds with Joe's 'impotence problem' where Freud has failed ('I spent six months in Vienna with Professor Freud, flat on my back,' he moans. 'Have you tried American women?' Sugar innocently asks). At the same time is revealed Monroe's fragility and poignancy. The song 'I'm Through With Love' not only exposes Sugar's immediate feelings; in retrospect, it encapsulates a tawdry childhood, three disappointing marriages, the adulation/mockery of curious fans, even a final, abortive 'phone call. Jack Lemmon remarked: 'It *was* moving. Acting with her on that picture, I felt that she was, inside, deeply unhappy. She was no "giddy blonde". She had a certain intelligence in and about her work, and she was smart enough to use herself to make Sugar come alive.' The character of Sugar Kane serves as evidence that, of all the Monroe impersonators, there was none so skilful as Marilyn Monroe.

Monroe was now deemed too successful a star to stray outside of her type, and too unpredictable a person to trust with much power. As a movie star it was unthinkable to allow Monroe to die before the end of her films; as a person it was, it seems, sadly easy to allow her to die prematurely. Her career was buoyed by her celebrity as she weathered the critical storms after the exceedingly poor *Let's Make Love* (1960). In 1961, her marriage falling apart and her health a serious cause for concern, she made her last film *The Misfits*. This film has so many important resonances that it will be treated in depth in chapter 6. Significantly, in the light of Monroe's Method, is the movie's apparent concern for the authentic: a 'real' Western location, a nagging interest in fundamental 'American' issues, and a sense of each actor's personal investment in their own character's identity and destiny. Her most notable scenes are those when Roslyn expresses her horror at the inhumanity of the men to the mustangs they are catching. Director John Huston remarked: 'She went right down into her own personal experience for everything, reached down and pulled something out of herself that was unique and extraordinary. She had no techniques. It was all truth, it was only Marilyn. But it was Marilyn plus. She found things, found things about womankind in herself' (in Summers 1985: 196).

Monroe did begin another movie, directed by George Cukor, called *Something's Got to Give* (1962). She was on a salary of $100,000, whilst Elizabeth Taylor was working for over $1,000,000 on *Cleopatra;* the latter drained the studio of capital, and Monroe's movie was subjected to close financial scrutiny. She became ill, rarely turned up, and was fired. Several weeks later she was dead. The brief sequences that she had shot for the movie were released with the Fox 'tribute', *Marilyn* (1963); they show little trace of the round-faced pin-up of the early days, but instead a person of grace, luminosity and awful fragility. In a series of costume tests she walks and walks again, in silence, and the hieratic, ritual magic of the rushes is haunting and unforgettable.

After her death began the re-runs, revisions and revivals. Early movies were re-released, out-takes were taken back in, prints were isolated, parts were analysed, the career was 'appreciated', and audiences returned to her movies and followed and felt for her with a poignant sense of nostalgia. Video releases allowed viewers to freeze the frame and look, or rewind and look once more. The movie Monroe is now massively overdetermined, folding in and out of our memory of 'Marilyn'. To speak of our 'involvement' in movies is too modest a term: we involve the movies in us, they become further fragments of what happens to you and to me and to them, further cards in the shuffle of memory. Monroe's performances were prepared, thus planned, yet they still provide for us an element of surprise. It was always part of the grain of film that, however studied the lines and strict the direction, the movement of the actor was essentially improvised. Such everyday actions as walking across a room, lifting a glass, sitting down, looking around, adding a thought to a conversation – they could all go one way or another. Monroe's resources were given, but their application to each new crossroads was an improvisation of meaning, out of the present.

As with all myths, our memory of Monroe is strong in generalities and weak in details. Did she make that memorable quip in *Bus Stop* or *Some Like It Hot?* Whom did she marry in *How to Marry a Millionaire?* Each person carries a *montage* of memories taken from a series of Monroe movies, a kind of tape-loop of recurring themes and images. The *oeuvre* encourages this 'overspill', this overlapping of motifs from one film to the next, so that 'Marilyn' remains after the individual films have faded and flaked away. There is an unusual continuity of character from one movie to the next: secretary, showgirl or sex-pot. The 'look' is frequently the same: blonde, revealing, sexually significant. The character often plays upon the breathy voice, quivering lips and the comic quips, and betrays a habitual weakness for 'father' figures: 'Daddy' *(Ladies of the Chorus, Gentlemen Prefer Blondes, Some Like It Hot, Let's Make Love)* and 'Uncle' *(The Asphalt Jungle, Don't Bother to Knock)*. She carries a typically frivolous catchphrase: 'Thank you ever so ' *(Gentlemen Prefer Blondes)*, 'just creamy' *(How to Marry a Millionaire)*, and 'just elegant' *(The Seven Year Itch)*. In both *The Seven Year Itch* and *Some Like It Hot*, Monroe's character confesses that certain tunes ('chopsticks' and 'Come to Me My Melancholy Baby' respectively) make her 'go goosepimply all over!' She is in constant search for her 'ideal man': 'kind and generous' *(Gentlemen)*, 'quiet and gentle' *(The Seven Year Itch, The Prince and the Showgirl, Let's Make Love)* 'shy and intelligent' *(Some Like It Hot)*, and typically summed up in *Bus Stop:*

> I want a guy to look up to and admire, but I don't want him to browbeat me. . . I want a guy who'll be sweet with me, but I don't want him to baby me. . . I'm going to marry a man who will have some regard for me apart from all that love and stuff.

Monroe is a woman constantly surprised by isolated instances of male civility: 'He called me a lady!' *(Full House)*, 'You took your hat off to me!'

(The Misfits). Perhaps even more hypnotic are the interweaving references to Monroe's extra-filmic identity: the edition of *U.S. Camera* featuring the prize pose from 'the Girl' had also featured Marilyn Monroe; when the man is asked who the girl in his kitchen is, he replies, 'Wouldn't you like to know? Maybe it's Marilyn Monroe!' *(The Seven Year Itch)*; Gay's wardrobe door is opened to reveal a selection of Marilyn Monroe pinups, much to the embarrassment of Roslyn *(The Misfits)*. The ideas and images roll on and over each other. To know one film one needs to know them all, and to know them all one needs intimate knowledge of one, and still, it seems, we do not quite capture the 'real' Monroe. As Benjamin says, no single image can satisfy us, no particular performance can be pointed to with authority, for with the fan of memory 'only in its folds does the truth reside'.

Monroe repeatedly represented on film a woman with an innocent, almost infantile quality in her matter-of-fact attitude about her sexuality. She was seen to strive to resist being hurt, or to hurt others, and she uses that sexuality to give pleasure rather than to cause pain. Yet, in the narratives, these sexual relationships are unmarked by emotional attachments and hence she is defined by sheer physicality. Her exploration and exhibition of her gender was positioned by her directors in order to make her 'every man's love affair', encouraging the kind of masculine metaphors manufactured by Mailer (1973: 15) 'The sexual immanence of her face came up on the screen like a sweet peach bursting before one's eyes.' The sheer scale of this screen spectacle made the body a landscape, discussed by J. G. Ballard in *Love and Napalm*: 'Marilyn's pitted skin, breasts of carved pumice, volcanic thighs, a face of ash. The widowed bride of Vesuvius.' The complexity of response to the movie Marilyn Monroe does not reside simply in sense or sensibility, scale or sexuality: it does turn upon her projected body and its paces, but equally upon the compassion that was lost and confusing with it, and upon a capacity for wit and laughter that many would give almost anything to elicit and satisfy. Billy Wilder now concludes:

> She was an absolute genius as a comic actress, with an extraordinary
> sense for comic dialogue. It was a God-given gift. Believe me, in the
> last fifteen years there were ten projects that came to me, and I'd start
> working on them and I'd think, "It's not going to work, it needs
> Marilyn Monroe". Nobody else is in that orbit; everyone else is
> earthbound by comparison. (In Summers 1985: 178)

Marilyn Monroe was a performer whose exceptional individuality clashed with a studio stereotype to produce a significant and superior image. Her body of films carry the pulsing air of incommunicability which may touch the edge of any experience and placement: the curve of the fingers or the wave of a white dress, a rounded mouth, the sudden rise of the body's frame as it is caught by the colour and scent of another, despair in a final parting, laughing mostly about nothing, the lover gone but somewhere now which begins from here – spools of history that have unwound only to us now, occasions which will not reach words for us now, and if not now, never. In the movies, we are not forbidden to cross the line between actor and incarnation, between action and passion, between profane and sacred realms. In the movies, the barrier to Marilyn Monroe is time.

Entertaining the troops in Korea: 'The happiest experience of my life.'
Photograph reproduced by permission of Aquarius, London.

5

Marilyn and the Public

We knew the stars as we know God. They were all-pervading but remote; we created them, but their fate was not our fate and they remained intractable, unlikely to respond to our prayers.

Quentin Crisp

With fame, you know, you can read about yourself, somebody else's ideas about you, but what's important is how you feel about yourself – for survival and living day to day with what comes up.

Marilyn Monroe

Marilyn Monroe was pre-eminent in a Hollywood which boasted having 'more stars than in heaven'. 'I didn't look at myself as a commodity,' she said at the end of her life, 'but I'm sure a lot of people have. Including, well, one corporation which shall remain nameless' (Meryman 1962). As the last great star of the studio era, Monroe was presented to the movie-going public as the epitome of glamour, a celebrity for all seasons and all cinemas. Her studio saw her as a highly marketable symbol of sex and stardom and American success, a glamorous fan-fare for public consumption. Yet what relationship did Monroe have with her public? As a sex symbol she was a tantalizing token from an industry which 'perpetually cheats its consumers of what it perpetually promises', a market wherein 'the diner must be satisfied with the menu' (Horkheimer and Adorno 1982: 363). There was something profoundly contradictory about the popular image of Monroe: omnipresent yet unapproachable, the individual struggled to make her public sympathize with the self behind the celebrity. It is this tension, this ambiguity, that will be explored in what follows.

The word 'star' is a more beautiful and more significant word than the word 'actor'. Not every actor thrust centre-stage is a star, no matter how intensely she is publicized and popularized. In a modern era of mass communications and mass surveillance, we have become singularly interested in the qualities and curious subtleties of individual figures; never before have we made such spectacles of ourselves. The star is seen as the focal point wherein the personal and the public meet and mediate. The stars are simply to gaze at, after the fact, and their movements divine their projects. The star is both real and mythical, a person and a commodity. The star draws us to the cinema, where we can enjoy the presence of the image with sound and vision. Yet that walking, talking, performance is never quite sufficient, and in fact it serves to increase our curiosity about what the star is 'really' like (behind the scenes, beneath the make-up) hence, the cycle of fandom, scandal and gossip. Stardom's effect is awesome: coherent, marketed personalities can be purchased by a mass of people, and this coherence overrides our knowing that real personalities are unresolved. The confidence of stars has coaxed us into acting: we need to present ourselves, and can no longer simply be. Stardom is one more condition of knowing duplicity, a duplicity well-appreciated by Marilyn Monroe:

> I don't care about the critics. . . The only people I care about are the people in Times Square, across the street from the theater, who can't get close as I come in. If I had light make-up on, they'd never see me. This make-up is for them, so that when I wave to them it will soften out in the distance across the square. (In Taylor 1983)

We all know 'Marilyn'; we just do not know what this means. The problem was stated succinctly by Benjamin (1982: 245): 'The essentially distant object is the unapproachable one. Unapproachability is indeed a major quality of the cult image. True to its nature, it remains "distance, however close it may be". The closeness which one may gain from its subject matter does not impair the distance which it retains in appearance.' It is common for the photographed subject to gaze at you (i.e. at the lens). In contrast, in the cinema it is forbidden for the actor to look at the camera (i.e. at you). In the cinema the audience gaze at the star, the star never gazes at the audience: she gazes at everything *except* us. Monroe seemed close without touching, a bright star in some distant constellation. Her very public image served as a meeting point for a mass of people in a modern, manufactured milieu. The movie theatres were built in the most crowded parts of cities; they fitted into a pattern of life that included tenements and factories. The industry fastened on a profitable service to bring relief, adventure, pleasure and leisure to people tired and sometimes near to despair. The everyday monotony and the real obstacles to hopes were offset by this nocturnal expansiveness. The fascination flickered on during the daytime: fan magazines and the innuendo of the gossip column identified the devious pleasure of mistaking actual and functional identity, and picked out a public's unwholesome peeping curiosity about the immense screen celebrities.

We 'see' the movie star as we see any other kind of star: the 'author' of the attractive aura may have long-since perished, but the afterglow is for us the very present source of interest. Although cinema is a popular spectacle, a personal quality was inscribed by the spectator. The activity of sitting in the dark and 'believing in' the images is complex and delicate and demands a philosophical agility that few people find easy to articulate. Such an agility was afforded a persistent persuasion by the panoply of commentaries which wrapped themselves around the impressions accrued from the passing screen themes. We seem to repossess past time when we watch cinema stars. The identical, but repeatable, rendering of duration is one of the most beguiling and bemusing of film's capacities. Stars are ascribed a lustre of nostalgia, and the movies seem a most peculiar annex to memory. Memory is screened by the movie: we are having to begin to remember as we see. The star succeeds herself. The most satisfactory movie experiences lead to one returning home, half-recollecting, half-reinventing that face and those moments from the movie. The fan takes away a seed of the star. We all cultivate our conceptions of Monroe.

Stars, whether their image has direct or indirect political implications, have the same prominence in public culture as political functionaries, but this prominence is as a result of their personally based rather than collectively based accomplishments. Monroe, it was said, was a 'singularly attractive' phenomenon. At the level of film (and Monroe was, primarily, a film star) what is involved is the question of the extent to which stars are prepared either to enhance the impact of the narrative by submerging themselves in character or alternatively to enhance the impact of their image. However, it is the state of economic organization, and not solely the intention of the stars, that makes these strategies dominant or subordinate options. It was Monroe's campaign against her stereotype that brought into relief the role of her studio in the development and determination of her public image.

The impact of audience selection on Hollywood was a complex process of interaction, marked by a large degree of inertia, channelling and predetermination (see Handel 1950). The general point that emerges from such an arrangement is that it is entirely consistent with industrial survival and predominance that the public influence over film production should be restricted to the selection of stars from a predetermined list of contenders. The star system was the form of competition between the major studios that was consistent with the stabilization of monopoly control. Forms of competition that would widen the range of choice available for public selection were incompatible with a producer- rather than an audience-centred system of production. The 'stock' of stars was always in need of revision: a few years were the most the majority would expect as a saleable commodity. The old head of Columbia Pictures, Harry Cohn, had devised a foolproof method for detecting 'sexual chemistry' on the screen. At a preview he would sit down in the darkened theatre, think to himself of his endless need for new stars, and wait for a feeling in what was euphemistically known as his 'butt'. If it tingled, that was good – the stars were out tonight. If he just wriggled around and got 'kinda sore', that was bad. Outside the mystical realms of movie methodology there are more reliable ways of studying 'star quality'. Cohn and his colleagues used them to their advantage.

As was noted in chapter 1, the star image is a construction composed of a number of media strategies which include promotion, publicity, films and critical commentaries. Promotion consists of studio-monitored press releases, fan club reports and public appearances. Publicity is what the press is able to 'find out' from gossip and interviews. The films themselves include the presence of the star, her image to some extent mediated by the previous two sources of information and influence. After the film and before the next one, critical commentaries in newspaper columns, reviews, books and biographical analyses serve to stimulate further shifts in the public perception of the stars. This circuit of production, distribution, exhibition and reception is continuous and cumulative. The star image is a 'textured polysemy', having several meanings and effects, several levels of signification. The potential for conflict between these meanings and levels is one of the key aspects of the image. Through time, Marilyn Monroe appeared as a starlet, a sex symbol, a succession of screen characters, and as a partner of famous men and a victim of several infamous plots. Within particular periods of her career, her image was the site of a struggle for meaning between studios, screenwriters, press agents, promoters, and Monroe herself. Throughout, the public followed the star and her successes and struggles.

Powdermaker (1950) observed that the relationship of 'fans to their stars is not limited to seeing them in movies, any more than primitive people's relationship to their totemic heroes is limited to hearing a myth told occasionally'. In the case of Marilyn Monroe, the public followed the trace of her life as it acquired a higher profile and a wider currency. In the movies, Monroe was practically 'present' to the fans. The basic psychological machinery through which most people relate to film involves some combination of identification and projection. The whole character of the audience situation encourages high levels of immersion: a condition of nearly total darkness, dominated by the proximity of a large, projected, sound-synchronized, moving image. With 'distraction' minimized, the audience is almost compelled to enter the world of the movie. There are different points at which the identification-projection pattern may apply. Two points stand out from the rest: the performer and the genre. In the days of the 'mass audience' the performer was the focus; the star system provided the basic leverage for audience involvement. From the fifties, along with a clearer emergence of 'subcultures of taste', genre increased in importance. Monroe's career crossed the two trends, and in any case the two are by no means independent. Monroe developed an affinity for particular genres (primarily musical-comedies).

Hollywood, recognizing the relatively fixed audience conception of the 'Marilyn' persona (and, to some extent, consciously creating it), could build on it as an element in the communication process. At its crudest this led to the star vehicle, such as *The Seven Year Itch* – a movie created primarily as a frame within which Monroe could be profitably exhibited; but more subtle uses blended Monroe into the overall pattern of the film, such as *Some Like It Hot*. The studio thus recognized that the relatively fixed persona of 'Marilyn', created through the movies themselves and through the publicity process, was a central element in audience involvement. As a young woman fan of Monroe recalls, the 'fixed' image did not prevent numerous revisions from each spectator:

One of the loveliest things about her was that she was accessible to women. The culture kept telling us she was the ultimate sex symbol, the blonde bombshell. And she was still accessible to women! I never felt excluded by her, not when I was little and she was my Magic Mommy, not when I was older and she was some sort of a sister. (In Oppenheimer 1981: 37)

The stable 'Marilyn' image was thus open to various uses and interpretations by the audience. The fans could identify with Monroe in a range of different story circumstances, and project their designs and frustrations into this intimately accepted character; it was this that gave rise to the familiar finding that audiences are drawn to stars of their own sex (Handel 1950). In Monroe's case, whilst women may have felt some emotional affinity for her, the men seem to have been identifying themselves with Monroe's screen 'partner'.

The 'mixed' movie audience is significantly influenced by a number of extra-cinematic pressures. In what is an underdeveloped area of research, there are fascinating glimpses of a movie audience in which men and women watch and converse as individuals, as friends, as partners, as critics – a multiplicity of roles, merging into one another, sometimes contradicting one another. An example is found in Dennis, Henriques and Slaughter's (1964) record of the impact Monroe's *femme fatale* from *Niagara* had upon an audience of miners and their wives in the North-East of England. The study observes (p. 216):

> In the bookie's office or at the pit they made jokes about the
> suggestiveness of Miss Monroe, about her possible effect on certain
> persons present, and about her nickname, "The Body". Indeed, any
> man seemed to gain something in stature and recognition if he could
> contribute some lewd remark to the conversation. On the other hand,
> in private conversation with a stranger the same men would suggest
> that the film was at best rather silly, and at worst on the verge of
> disgusting. Finally, the men's comments in the presence of women were
> entirely different. In a group of married couples who all knew each
> other well, the women said that they thought Miss Monroe silly and
> her characteristics overdone; the man said that they liked the thought
> of a night in bed with her. The more forward of the women soon
> showed up their husbands by coming back with some remark as "You
> wouldn't be so much bloody good to her anyway!" and the man would
> feel awkward.

This example shows how the movie experience – especially the attempt to *share* it – can bring into play the tensions within each person's cluster of social roles. The disjunction between private fantasy and public pronouncement may become marked: the man may feel a need to impress his male companions by making a vulgar comment about Monroe ('some lewd remark'), whilst fearing seeming a fan of 'low' culture to the discerning stranger ('rather silly', 'on the verge of disgusting'), whilst sensing an opportunity for masculine arrogance and self-celebration in the presence of women friends ('a night in bed with her'). Thus, the movie memories can serve as source material for the enforcement or undermining of one's social identity and status.

The cinema helped to legitimize the practice of voyeurism on a social scale, drawing a discreet veil of darkness over the deleterious aspects of such activity. Identification with the screen star sometimes passes over into imitation (witness John Hinkley's pathological involvement with *Taxi Driver*). Monroe inspired many models and mimics. A typical example of imitation at the time shows the ways in which the star's influence saturates the fan's everyday behaviour; a young woman says,

The settings of the love scenes always hold my interest and I've always
noted little tricks (which I've put into practice) such as curling my
boy-friend's hair in my fingers or stroking his face exactly as I've seen
my screen favourites do in their love scenes. One of the first things I
noticed was that an actress always closes her eyes when being kissed
and I don't need to add that I copied that too. (In Mayer 1948: 42)

Thus, the star system was a central factor in the psychology of the
movie-audience. Women such as the above example were given an array
of socially certified ideal-types which served as ideas to strive for and
images to aspire to. Monroe represented such an ideal for many women
viewers, but she also elicited emotional identification from homosexual
male viewers – particularly when her 'psychological problems' were more
well-known. Arthur Bell, a representative of the gay sensibility for the
Village Voice readership, explains:

If you feel that your true nature is female and you are gay, then you
have the persecution of being gay. And when you see a larger-than-life
that's gone through a decline, it becomes kind of a nasty Valentine to
the world: *see what you did to me*. I think that's true about most
actresses that gay males idealize. I don't think they'd be that interested
in a complete, adjusted woman like, say, Simone Signoret. (In
Oppenheimer 1981: 50)

The typical masculine response is the most vocal and the most
predictable: the epithets were frequently culinary, ranging from 'tasty' to
'good enough to eat' to 'a gorgeous dish, a real feast for the eyes' – every
man's love affair with 'Mom's apple pie'. Perhaps Billy Wilder gives the
important additional point about the masculine perception of Monroe
when he stresses her 'playful' quality. 'If I told my wife after I worked
with Monroe, "My God, I'm crazy about her", she says she would
understand it. But if I told her, "My God, do I wish I could spend a night
with Farrah Fawcett-Majors", she would hit me in the mouth' (in Taylor
1984).

It is important to remember that 'going to the movies' is a social
institution as well as an individual experience. The relative homogeneity
of each group's response to Monroe was a conscious goal for her studio,
which used technology and publicity to foster this effect. In the cinema the
eye is directed and pleased by Hollywood imagery, and the same gentle
persuasion works in pin-up photography and graphic advertising. The
near-perfection of camera angling is the climax of meaningful shape
imposed by light and point of view. It does not simply discover things in
the image, but locks spectator and spectacle together in a process that

whispers, 'This is insight.' The Hollywood message is tacitly commercial, and the text is its own packaging. Although Monroe was seen by the public through the prismatic 'Marilyn' persona, the match was by no means perfect, and this only added to the intrigue.

On the screen, the star was sacred, beyond reach, out of earshot; a physical presence was another dream. A relatively direct form of promotion saw Monroe sent out to 'meet the people'. The producer of *Love Happy* (1949) used the new starlet Marilyn Monroe to promote the movie. He pumped up some fresh publicity, portraying her as Hollywood's very own 'orphan' struggling to succeed, and sent her off on a nationwide promotion tour. In New York State, Monroe's key appearance was outside *Photography* magazine's 'Dream House', a publicity caper simultaneously promoting *Love Happy* and household products. At the urge of studio publicity men, showbusiness columnist Earl Wilson, interviewing the unknown Monroe, introduced her as the 'Mmmmmm Girl'. In Manhattan she was escorted to the exclusive El Morocco nightclub, and was photographed being entertained by eminent dress manufacturers and well-known film figures. From New York to the Midwest, and Monroe was made to pose as 'the hottest thing in bathing suits cooling off again'. The publicity generated by the tour was a useful catalyst for Monroe's own emergence as a 'personality'. Later, such forms of public appearance could occur less frequently and more memorably.

In April 1954, Monroe and her new husband left on an extended wedding trip to Japan. When the plane touched down at Tokyo's international airport, the couple were disturbed by the excitement their appearance caused. Hordes of Japanese men and women surrounded them, screaming *monchan* ('precious little girl'). Some fans threw flowers, but the bolder ones grasped at Monroe's hair, pulling some of it out by the roots. Whilst DiMaggio made his own celebrity appearances, Monroe accepted an invitation to visit the American troops in Korea. Via Seoul, Monroe was flown by helicopter towards the encampments of the 1st Marine Division and the 45th Division of the Army. The men were crowded together in a crude amphitheatre, scooped out of a hillside. Monroe changed into a figure-clinging gown of plum-coloured sequins, cut typically low and exposing her nearly bare chest to the freezing winds. She was resplendent in rhinestones to go with her opening song, 'Diamonds are a Girl's Best Friend'. In the middle of the song, she stopped and walked over to a soldier in the wings who was about to snap her picture. She leaned towards him and gently plucked a lens cover from his camera, saying: 'Honey, you forgot to take it off'. She finished the song and went into Gershwin's 'Do It Again'. The colonel took her aside and ordered her to sing more 'clean' numbers next time. The public performance was a glorious one that would prompt Monroe to repeatedly describe it as the highlight of her life. She had certainly never experienced such an outpouring of affection and regard before, nothing as concentrated or as climactic. All of her life she had been seeking just the kind of total acceptance she was now receiving and the world press was covering. It also inspired an increase in those fans who followed her on a day-to-day basis.

'The Monroe Six' consisted of young, devoted fans who seemed to onlookers to have no homes to go to. They lived for fleeting glimpses of Monroe as she walked her dog, or rushed to get inside a taxi, with a wave to them or a kiss blown their way. The vigil began when Monroe moved to New York and it continued for the rest of her life. John Reilly had followed her since her notorious location trip to Manhattan for *The Seven Year Itch;* Jimmy and Eileen Collins were brother and sister united in their devotion to 'Marilyn'; Gloria Milone, Freda Hull, and Edith Pitts were teenagers who were determined to follow Monroe to every function. She warmed to her special group of young fans, inviting the Monroe Six to Connecticut for summer picnics. A particularly vivid and interesting account by a fan of a meeting with Monroe comes from James Haspiel. It is sufficiently revealing to be quoted at length:

> I was sixteen years old, from the West Side... I was interested in
> movie celebrities as something to do, instead of going around like the
> other guys heisting cars. Marilyn was as a personality a kind of escape
> for me... I heard on the radio she had come into Idlewild and gone to
> the St. Regis Hotel. That night I went to the St. Regis. There were two
> entrances. It turned out I was waiting at the wrong one, for I heard a
> cry from the crowd and ran round. I saw nothing but throngs of
> people near the entrance. I guess I expected something tall and
> wonderful, like on the screen, someone I couldn't miss. I couldn't see
> Marilyn for the people around her. I pushed my way through and
> reached the centre and found a tiny girl in a full skirt, the kind that
> stands out, a sweater, wearing cuban heels which made her even
> shorter. She worked her way through the crowd and said "I'll be back
> in two hours". I was back... and she was getting out of a cab. I didn't
> have a camera... I pushed my way through to Marilyn, I said: "Miss
> Monroe, would you please kiss me?" She hesitated a moment, smiling,
> "Just on the cheek", I added. She turned me to the side and kissed me
> on the cheek. I ran home and woke everybody up in the apartment and
> told them Marilyn Monroe had kissed me. (In Guiles 1985: 260–1)

Not only was Monroe conscious of her effect on these fans, she also took an evident pleasure in 'meeting the public' in an unpretentious manner. W. J. Weatherby (1976: 207) recalls sharing a New York taxi cab with

Monroe when the driver requested, and received, her autograph: 'He was shy with her but familiar, too, as if he had a personal connection. A lot of her fans seemed to feel that way. Marilyn was one of them.' This was always the intended image of stars such as Monroe, yet in her particular case there seems good reason for believing it to be true.

At the start of the 1950s, it was estimated that Hollywood studios received over a quarter of a million fan letters per month. A major star would have expected to receive about three thousand letters per week. *Time* (11 August 1952) reported of Monroe: 'She currently gets more than 5,000 letters a week from smitten admirers. Soldiers in the Aleutians voted her "the girl most likely to thaw out Alaska". A whole U.S. battalion in Korea recently volunteered to marry her. Students of the 7th Division Medical Corps unanimously elected her the girl they would most like to examine.' The studios kept a careful watch on the popularity of contract stars such as Monroe, actually weighing the amount of fan mail she received each week. Furthermore detailed feedback on a star's public popularity was sought in several institutionalized ways. Exhibitor opinions were voiced during conventions, in trade papers, and in an ample and diligently supervised correspondence with the motion picture companies. Exhibitor enthusiasm was eagerly courted in order to ensure a good national publicity and advertising campaign. The studios conducted their own audience investigations, often using questionnaires which typically featured the 'want-to-see' question: 'Would you want to see a picture if you knew nothing about it except that Marilyn Monroe is playing a leading part in it?' Such a question was posed in the light of previous roles played by Monroe in particular kinds of movies, and probably contributed to her studio's determination to keep her playing variations on her successful 'dumb blonde' role. Another typical question from a studio study (see Handel 1950) was:

Question: What do you Think about Marilyn Monroe as 'Cherie'?
Excellent
Very good (looking), like her very much
Good (looking), like her
Fair, pretty good
Didn't like her much, not good
Did not like her, bad
Other

The percentage listed male preferences, female preferences, and the overall total. The relatively peripheral role played by Monroe in *The Asphalt Jungle* was deemed too minor to merit her a screen credit when the movie was first previewed. However, audience reaction was very positive, and many people asked on their preview comment cards who the 'young blonde' was. Thus, Monroe's name was added to the credits.

The popularity of movie stars, and relative booms and slumps in their popularity, were matters of great concern to the studio to which they belonged, to the stars themselves, to their agents, and to numerous other people active in production, distribution, and exhibition. 'How was I?' became a crucial question amongst stars and star-makers during the studio system. (When records became more freely available we will be able to reconstruct a clearer picture of the star–audience relationship.) Audience response was sought by the studios in order to monitor movies and movie stars. Preview screenings offered a 'trial' for stories and star quality, and views were recorded. Star-ratings attempted to provide data pertaining to the relative popularity of the stars covered by studio-inspired surveys. The effect of different roles and genres on a star's popularity (serving to indicate the kinds of parts in which the audience liked or disliked the star); special sex, age, social, and geographical breakdowns to determine audience likes and dislikes of various strata: practically every permutation was explored. One may well warm to Cohn's 'tingling butt' technique when confronted with the mass of information and audience surveillance encouraged by the studios. 'Fashion', Quentin Crisp remarked, 'is *instead* of style': the studio surveys treated audience reports with such reverence that few stars could hope to radically change the direction of their careers – the audience liked them as they were, and it seemed they wanted more of the same.

The indexes could be established in several ways. One method favoured by the Motion Picture Research Bureau was to show audience members the name of a star and to ask them to indicate their degree of interest with the aid of a checklist. The different evaluations ranged from 'Marilyn Monroe is my favourite movie star,' to 'I dislike Marilyn Monroe very much.' Each of the various evaluations corresponded to a certain 'appeal index rating', ranging from 100 to zero. The average computed from the different reactions was the popularity index. Another method expressed the popularity index in terms of 'want-to-see'. The proportion of affirmative answers obtained to this question was the basis for the popularity index. Audience Research, Inc., contributed to a more sophisticated version of star ratings by providing breakdowns of the total ratings according to gender, age, income, and frequency of attendance. With this information the movie producer could avoid combining stars whose following came from the same stratum of the audience, and could compose a cast which would appeal to many different sections. This was the kind of approach untilized in the making of *Gentlemen Prefer*

Blondes. Thus, the relationship between the cinema and the spectator was a two-way process, with the audience having an effective, albeit manipulated, role to play on future productions.

There is a very important social structure attached to the institution of movie-going. This surrounds the role that Lazarsfeld and his colleagues christened the 'opinion leader' (see Katz and Lazarsfeld 1955). In Hollywood, two of the key opinion leaders were Hedda Hopper and Louella Parsons, both with widely syndicated columns which for a number of years assumed a quite ludicrous importance for such ill-qualified critics. Hopper and Parsons were certainly instrumental in Monroe's success. Parsons, in particular, performed an invaluable service for the young starlet: shortly before her breakthrough, Monroe was chosen as Parsons' 'No. 1 Movie Glamor Girl' – a selection that attracted heavy newspaper coverage. 'She had to be the winner,' Parsons wrote. 'It is Marilyn who is number one on all the GI polls of Hollywood favorites, and number one on exhibitors' polls of box office favorites. She is the number one cover girl of the year, and certainly number one in public interest wherever she goes.' Monroe showed a significant measure of foresight and singlemindedness in her early cultivation of press contacts. She seems to have been able to develop friendships with columnists which continued throughout her career.

The centre of printed reflections on Monroe was the magazine, oriented towards the youth market. This medium—glossy, gossipy, keen and confessional – elaborated the myth of celebrity. Fan magazines *mirrored* the reader's values: 'they are fascinated, I am fascinated.' Fan magazines maintained and sometimes *manufactured* this fascination – through 'favorites' polls, through requests for feedback ('this is *your* magazine'), and through special merchandise. Stardom is an image of the way stars live. Commonly, this generalized lifestyle is the assumed backdrop for the specific personality of the star and the details and events of her life. As it combines the extraordinary with the everyday, and is seen as an articulation of basic Western values, there is little conflict here between the general lifestyle and the particularities of the star. In certain cases, however, the relationship between the two may be ambivalent or problematic. The later Monroe's aspiration to the condition of stardom and her melancholy on attaining it are part of the pathetic/tragic side of her image. This aspect of the image was focused upon by the many feature articles on 'Marilyn's men trouble'. A central theme in all the fan magazines was love. This was achieved partly by the suppression of film-making as work and partly by the overriding sense of a world wherein material problems have been settled and all that is left are 'relationships'. What is particularly interesting about these magazines is that love is often not so much celebrated as agonized over. What emerges especially strongly in the stories is a concern with the *problems* of love. Articles with titles such as 'Tarzan seeks divorce' and 'This Year's Love Market' predominate (see Bego 1986).

Movie studios twisted the facts about the star's background in order to create a manufactured delusion. Often these Hollywood idols were in reality quite the opposite of the personalities they seemed to be in the press releases or on the silver screen. 'Exotic vamp' Theda Bara was the movie Cleopatra from the 'magical East'. In fact she was Teodosia Goodman from Chillicothe, Ohio. 'Cute-as-a-button' flapper Clara Bow had her career cut short when it was discovered that she had spent the night with the entire University of Southern California football team. Marilyn Monroe was portrayed as the 'poor orphan girl' who is suddenly 'getting a break'. It was in the context of these fan magazines that the 'cute', childlike quality of the 'Marilyn' image was especially evident. She was found advertising lipstick, hair gel, stockings and soap. *Picturegoer* (10 October 1953) contained an advertisement for Hilton hair-dye which announced: 'Marilyn Monroe says Gentlemen Prefer Blondes. . . and blondes prefer Hilton.' For several years it seemed that truffles were Monroe's best friend: she was pictured with them by the bath, on the bed, and even backstage. She appeared in a variety of 'fun' poses: the studio planted a 'criticism' of Monroe's expensive dresses, and then planted a reply which argued that she would 'look neat' even in a potato sack. Monroe was duly pictured, looking duly neat, in a potato sack.

Studio publicists sought to establish Monroe as the 'ideal playmate'. The nature of the image was a fairly predictable outgrowth of her previous experience as a model and cover girl of 'girlie' magazines. One story reported:

> A loud sustained wolf whistle has risen from the nation's barbershops and garages because of Marilyn's now historic calendar pose, in which she lies nude on a strip of crumpled red velvet. Uneasy studio executives begged her . . . to deny the story. But Marilyn believes in doing what comes naturally. She admitted she posed for the picture back in 1949 to pay her overdue rent. Soon she was wading in more fan letters than ever. Asked if she really had nothing on in the photograph, Marilyn, her blue eyes wide, purred: "I had the radio on". (*Time* 11 August 1952)

The facts were never dependable: it was some time before her fan letters increased, she had grey eyes not blue, and her witticism was uttered with more of a stutter than a purr. Nonetheless, in addition to the overt sexuality (the 'wolf-whistle-from-the-barbershop' syndrome), three elements were significant: *hardship* (she posed to pay her bills): *dumb humour* (the 'purred' reply); and *innocence* ('doing what comes naturally'). Each of these three themes encapsulates an aspect of her public image. Monroe reflected on this episode when she was interviewed by Garson Kanin (1975: 286): 'all of a sudden, even those dumbheads at the studio began to see that it wasn't such bad publicity. It was sort of good publicity because it was kind of sassy. So that's when they stuck me right across the top of the whole *Niagara* billboard and, I don't know, things sort of got going from there.'

Once the persona was positioned and publicized, it could be pursued and pinned down. What the pin-up offers is magical access to the body of someone who is not really there. The pin-up proffers irresponsible access to a body that is, in effect, a decorticate preparation: a body that will go through the moves of sexual response, but whose movements are uninformed by reflective intelligence or sexuality. Monroe's 'nude calendar pose' was a famous instance of such a process. Picturing is not exclusively an activity for professionals: metaphorically, we all do it, envisaging, epitomizing the world around us and our relationship to it. To this extent the relations of author to model, of both to the image, and of all three to the spectator, serve as a format for a large proportion of the everyday thinking we all do. *Coronet* magazine, during 1953, explained that, 'the smart woman will keep herself desirable. It is her duty to herself to be feminine and desirable at all times in the eyes of the opposite sex.' The fan magazine positioned the 'glamour girl' in a space reserved for consummate conceptions. A meeting of a real person (you), and another real person (Marilyn Monroe), who has been turned into an object that turns back into a person in the mind's eye; a vision of flesh and blood that is both magically accessible and yet, in principle, beyond reach. For the spectator, Monroe was a recruit to that cast of characters who people one's daydreams, reveries and sexual fantasies; for Monroe, the spectator was likewise a fantasy: a member of the audience that silently looks.

Twentieth Century-Fox early on issued a series of pin-up pictures which presented Monroe's anatomical properties to the best advantage. She was then publicized by means of a succession of honorary titles inspired by the studio – such as the 'Woo Woo Girl', 'Miss Sexy Eyes', and the 'Blonde Belle'. Prior to her achieving feature-billing in a movie, Monroe was featured by *Esquire* and *Coronet* as well as a host of lesser magazines with a decidedly male slant (such as *Peek* and *Sunbathing Review*). In October 1952, *Coronet's* Grady Johnson wrote:

Prodded by protectors of public morals, Hollywood for twenty years has been telling the world, with some traces of truth, that its residents

were home-loving church-going folks. Belaboring the point, its
publicity made glamor girls out as drudges with housemaid's knee,'
whipping up an angel food cake quicker than you could say
censorship-is-ruining-the-movies. Then along came Marilyn Monroe.

A year later the mock-shock of the columnists continued:

> High-and-low sums up the latest trouble in Hollywood. And it's all
> over – did you guess? – Marilyn Monroe. . . [T]he women's clubs . . .
> have made it clear to Miss Monroe's boss, Darryl Zanuck, that
> Hollywood is in for another purity campaign unless something is done
> to curb the present spate of suggestiveness in films and publicity.
> (Reported in the *Daily Sketch* 26 February 1953)

An early Monroe story in *Collier's*, planted by studio publicists,
contained all of the principal thematic associations later integral to the
popular image. As the circumstances of her birth, her childhood trials,
and her early marriage were revealed, Monroe became all the more
provocative as a sex symbol. The absence of family, it seemed, made
Monroe appear more attainable.

Monroe's marriage to Joe DiMaggio made her studio issue several
careful press releases designed to balance the star's newly found bliss with
a persistent sense of her still being an idol in isolation. DiMaggio's
scrupulous avoidance of publicity facilitated the studio's attempt to focus
on Monroe. The marriage stories thus devote most of their space to the
description of 'Marilyn Monroe's Honeymoon'. *Modern Screen* (April
1954) included a series of pictures featuring Monroe and DiMaggio about
to be married, entering their 'modest motel', and the very happy
'Manager E. B. Sharp and Wife'. Mr Sharp ('and wife') informed
Photoplay that DiMaggio's most impassioned honeymoon night request
was for a television set. Monroe was quoted as saying, 'We want six kids';
DiMaggio, we are told, 'guaranteed at least one'. In order to add some
background colour to the story, DiMaggio's older sister, Marie, was
contacted to explain Monroe's important 'breakfastime job'; 'She puts on
the coffee, although she usually doesn't drink any – Marilyn is a tea and
milk girl – and then she squeezes the oranges and cooks up the bacon and
eggs. She's really very handy in the kitchen.' In case the reader is in need of
further reassurance as to Monroe's man-maid skills, Marie adds: 'I knew
she was a good girl the first time Joe brought her up. Right away she was
helping with the dishes.'

The movie magazines provided an arena for star publicity to be
presented in a peculiarly intimate form. Studio press agents would rewrite
biographical accounts of the star in the first-person, affecting an

appearance of a 'fireside chat' between the personality and her public. Such articles sometimes carried the aura of a confession about them, an intimate and cosy insight into the star's personal feelings, and thus served as a unique mode of star–audience communication. A piece ostensibly written by Monroe, 'Curves versus Character', appeared in the *Hollywood Album* of 1953. One picture caption read: 'Marilyn steps out from under a shower after a heavy day before the cameras making *Gentlemen Prefer Blondes*. After exhausting dancing routines Marilyn finds a shower and a brisk rub down exhilarating and refreshing.' When Monroe gets to 'talk', the impression is of a star gossiping behind the backs of her publicists. Now, it seems, 'Marilyn' will relieve us of our fears of rejection and reassure us of her undiminished devotion:

> A lot has been written about what I do with my fan mail . . . that I
> strew it about the floor, then walk about over it in my bare feet. Silly,
> isn't it? I do with my fan mail exactly what it is intended for: I read it
> to learn how I'm doing. . . Fan letters can add more than a little to
> one's knowledge of human nature. That's something every actress must
> know a lot about if she's really going to build up the character she is
> to play.

The fan smiles in satisfaction, the knowledge that Monroe values her fans' help is a heartwarming thought. Monroe's 'report' concludes with an over-respectful explanation to her readers of her future plans:

> Now all I want to do is work – work hard for something more than
> wolf whistles. Don't think, though, that I'm not like every other girl. I
> would be very disappointed if men did not pay some attention. Every
> woman wants to have that male approbation, but an actress wants a
> little more – or, I should say, a lot more. She wants people to forget
> her figure, her walk, her clothes; she wants them to be so transported
> by her playing of a part, they come out thinking of the character.

After such magazine articles, such an ambition appeared still more remote and ill-conceived. Monroe's faith in her acting ability was constantly undermined by her own publicists, who stressed the importance of her physical appearance to her successful status.

The marketing of movies has always been a unique phenomenon, dealing with flickering shadows on a screen, the merchandising of memories. Sam Marx, in his study *Mayer and Thalberg* (1975: 14–15), remarks:

> Theatres are the stores where customers buy entertainment, but unlike
> most merchandising outlets, the buyer doesn't take his purchase away.

He pays for it and looks at it, then leaves with only the memory of it. When the last customer has paid and looked and left, the material that was bought still belongs to the man who sold it.

The fans of Monroe, before the next film-fix, could manage their addiction by forming a mutual association of fellow followers. Word-of-mouth publicity was a process encouraged by Hollywood. With the fan club, such a movement could grow into an institution. Memories of Monroe, culled from movies, magazines and momentary meetings, were cultivated in the company of like-minded people. The fans used the club as a source of group identity, a focus for their fantasies and a reinforcement of their fascination. The existence of the fan club collected Monroe's admirers in time and space, giving them a sense of solidarity. Information was exchanged and stored; attractions were charted; affections were dedicated and distributed *en masse:* the club acted as a container for interests, encouraging a *structured* appreciation of Monroe. Fans could tap a common memory bank, entering a force field of affects that converged on the solitary subject of Marilyn Monroe. Individual images were inscribed on a page like a kiss on a cheek and circulated amongst fellow fans. Memorabilia was produced and exchanged. Rebecca Beale, a young woman who became fascinated with Monroe in the 1980s, describes her room as 'a shrine', covered with 'Marilyn' images. She explained to me:

> I like to think that a part of Marilyn lives through me and I hope that a part of me belongs to her. . . I felt if I called to her she'd stop in the middle of a scene, look at me and say "Hello Becky". I felt that close. . . As I explored Marilyn I was also exploring myself. Maybe all this wanting to protect Marilyn is my own attempt at protecting myself. . . I feel I have reached the point now where all is left is to meet her, to be with her, because she's unveiled herself to me in every other possible way. This frightens me. (Correspondence with the author, July 1987)

The more obsessed of fans could even come to resemble, 're-assemble', their idol: the imitator could acquire the brand of stockings Monroe advertised, and her style of clothes, her make-up, hair-dye, possibly even her own old clothes – an elaborate 'repossession' of the body, usurpation of the figure by the fantasy. The process is rather like a fantom kiss blown to the fan from the other, 'a sort of cutting taken from one person and grafted on to the heart of another' (Proust). This 'Marilyn' is an occasion for the excitation of the pattern for desire – thought of her may call up feelings of love, melancholy, wonder, destructiveness, possessiveness, envy or longing.

Internationally, Monroe could still attract enough committed followers to merit *new* fan clubs being founded during the 1970s and '80s. Stewart Hutcheon, a British fan club organizer, tells me: 'I first took an interest in Marilyn Monroe several years ago when I first saw *Some Like It Hot*. I knew she was special, she wasn't cardboard . . . there was more to her.' He regards the Monroe Club as 'a forum for fans to talk about themselves and how they feel about Marilyn.' The UK Marilyn Monroe Fan Club provided personal testimonies from its own members for use in this study. The individual comments reflect the earlier concerns and convictions, whilst an additional feature is the fan's own interest in being an admirer of someone no longer alive. This interest is evidenced in both the search for *rarities* and the sense of *nostalgia*. As an example of the collector of the rare, a woman (aged 25) writes: 'I would like to own something of Marilyn's. I recently saw a bra of hers at an auction for sale at £500! But I'm a girl with limited funds, so I'll just have to dream.' The nostalgia aspect is explored by another female fan who explains:

> I've always loved Marilyn since I was very young. To me she was a beautiful woman in all respects, especially personality and looks. . . Marilyn was truly lovely and I'm sure she would have been to this day. . . It still amazes me that Marilyn has a cult following, and has had for about 35 years. With new books still coming out and no signs of the production line halting, it seems certain that Marilyn will be with us beyond the year 2000 – frozen in time, never ageing and always shining.

Potentially more disturbing are the complicated feelings of desire and resentment expressed by this (22-year-old) male fan: 'She's gorgeous – much more attractive than current stars like Madonna. I sometimes wish I could meet someone like her. . . I feel so frustrated – she's gone, we're never going to meet her. I don't know how she really died but it seems selfish – I mean we really care, even now.' This attitude is echoed in another (19-year-old) male admirer: 'You look at the poster and you think, she's such a good person. . . I wish I'd known her. . . I get moods when I try and forget her – it's just too painful.'

Rebecca Beale found a therapeutic method for dealing with these confused feelings: she wrote imaginary letters to 'Marilyn'. She tells me: 'I was really exposing myself. Looking back on it I don't understand properly why I felt like this' (correspondence with the author, August 1987). Watching *Some Like it Hot,* she became confused and angry, seeing Monroe 'so unknowing of my love for her, my devotion. I felt her slipping away from me.' The pictures were removed from the walls, the memorabilia put away – 'At that point I just wanted her out of my life.' The following letter (28 December 1986) was written:

Marilyn

I've always loved you, you know, from the moment I saw you. You exuded something so strong which really hit me. But now you've become fickle. How I wish you weren't. You've merely flitted in and out of my life. I would have given so much to you if you'd just let me in Marilyn. I didn't ask for much . . . I suppose you're gone forever.

Rebecca Beale explained to me why she then changed her mind and 'accepted Marilyn' again:

I realised how much she was part of my life. A happy part, a constant part. I could never live without her. So there I was putting all my pictures back up. Having her there brought back all the warmth to me. It was as though she was telling me, telling me she *does* belong to me. I still doubt and want but I've come to accept what we have now and how pointless it is asking for more.

Monroe herself was increasingly overshadowed by this 'Marilyn', this star construction. Movie stardom is a specialization based on conventional movie acting. Simply stated, acting is the process of consciously manipulating a self, which is occupationally defined as a personality implied by appearance, to project character. What is particular about the star is that she is not so much an actor but a self or personality that 'behaves'. To say that a star behaves does not mean that they are themselves, but rather that stars do not surrender the public personality to the demands of characterization. This distinction between actor and star caused Monroe, in particular, both professional and personal anxiety. Although she received the 'crystal star' from the French film industry and the David di Donatello statuette from the Italians, she felt slighted and belittled by the indifference amongst Anglo-American critics. In her 'private' life she suffered for her 'star profile'; she observed (Monroe 1974: 124):

I was becoming important enough to be attacked. Famous actresses took to denouncing me as a sure way of getting their names in the papers. . . In fact my popularity seemed almost entirely a masculine phenomenon. The women either pretended that I amused them or came right out, with no pretense, that I irritated them.

Her lovers, she said, expected 'sparks to fly' and 'bells to ring', and this contributed to her feelings of inadequacy. Stardom, for Monroe more than for any other, became an occupation:

I've always felt that the people ought to get their money's worth, and
that this is an obligation. I do have feelings some days when there are
some scenes with a lot of responsibility towards the meaning and I'll
wish, gee, if only I could have been a cleaning woman. I think all
actors go through this. We not only want to be good; we have to be.
(In Meryman 1962)

Monroe wanted to be a good actor, but her studio needed her as a great
star. She began 'wanting to be wonderful', and her stardom obliged her to
be wonderful everywhere, to everyone, every time. For her public
appearances, her studio ensured that she maintained a highly specific
image: false eyelashes, white gloves, and shoes dyed to match the colour
of her figure-hugging outfit. Even for casual 'at-home' picture spreads,
every hair on the head had to be immaculately in-place before a shot could
be snapped. Nothing was allowed to be spontaneous, nothing was left to
chance, all was premeditated in the service of media marketing.

When Monroe began her film career, the industry was spending over
$55.5 millions for advertising, in the USA alone (*Film Daily Year Book of
Motion Pictures* 1949). The number of advertisements placed daily in the
various media in America at the beginning of the 1950s was estimated at
15,700 insertions. When television began to break into the mass market,
the movies launched a rearguard action, struggling to preserve the public
demand for its product. The market for the star was becoming
increasingly difficult to manage. A number of factors contributed to the
audience's cultural context: the authority relations of opinion leadership;
the distribution of socio-economic characteristics; the 'simple'
distinguishing factors such as sex and age. Within these were the
variations in identification-projection that characterize the individual's
response to the movie-going situation. All of these, varying together and
separately, characterized the movie audience. The cinema offered a whole
entertainment package: one did not simply go to a Monroe movie, but
took part in a *social* event. Since the mid-1930s, exhibitors began to make
profit creation less dependent on the popularity of the movies: they
internalized the sale of a complementary good – the consumption of
popcorn, Coca-Cola and confectionery. Confectionery counters were
installed, and the audience was encouraged to move around. The
'intermission' was ritualized by theatre owners in order to provide a set
period in which snack food could be sold to a captive audience. Thus, the
exhibitors benefited from the social aspects of the cinema, and, in a way,
so did the consumer. Until well into the 1950s, movie theatres were one of
the few public institutions where the middle-class and poorer citizens of
the USA could indulge in cool, dehumidified comfort – the cinema
offered, at least, the relief of air-conditioning. Alongside the committed

movie buff, there were youngsters enjoying a shared evening out, families whispering to each other, passing the popcorn, drinking Coke, lonely viewers indulging in silent fantasies, and the odd 'petting' lovers. Although from some points of view the audience may be a 'mass', it is nonetheless also constituted by personalities, groups, classes and cultures. Not only does movie-making entail the coordination of a considerable number of artistic inputs in order to achieve a qualititative objective but, whatever the merit of that objective, it remains a scatter for subjective appraisal. This dual contingency has meant that no radical strategy of monopolizing talent can ever ensure a uniform run of compensating 'hit' movies.

Given such a level of uncertainty, such a policy would be damagingly expensive. By the same token, a strategy of competition through technical excellence would simply increase the cost of failure without necessarily increasing the number of hits. The corporate resolution to this problem – the problem of the costliness of monopolizing talent and expertise – was to develop a system of casual or picture-by-picture employment, which has affected both craft and artistic labour. In Monroe's era the stars were notable for the fact that of all the executive level salaried employees engaged on a picture-by-picture basis, they had the status of full-time salaried staff. Stars' salaries, like full-time executives', came from the fixed studio overhead charges, even though they were written into the cost structure of particular films. Further, the operation of the contract system meant that stars as a category of 'creative' personnel, let alone as individuals, were more closely tied to particular studios than other functionaries of the same rank (such as producers, directors, and writers), who tended to operate on a freelance basis, circulating between major and minor studios. Monroe soon made little secret of her contempt for this system:

> If you've noticed in Hollywood where millions and billions of dollars
> have been made, there aren't any kinds of monuments or museums –
> and I don't call putting your footprint in Grauman's Chinese Theater a
> monument – all right, this did mean a lot of sentimental ballyhoo to
> me at the time. Gee, nobody left anything behind, they took it, they
> grabbed it and they ran – the ones who made the billions of dollars,
> never the workers. (In Taylor 1983: 102)

Monroe used the power acquired from her star status in order to campaign in the press against the policies of her studio: 'I realized that just as I had once fought to get into the movies and become an actress, I would now have to fight to become myself and to be able to use my talents. If I didn't fight I would become a piece of merchandise to be sold off the movie pushcart' (in Taylor 1983: 100). She also said: 'An actress is not a machine, but they treat you like one. A money machine.'

The culmination of the 'Marilyn' stereotyping process coincided with Monroe's new commitment to independence and authenticity in her acting. The crisis occurred during the filming of *The Seven Year Itch,* a movie in which the Monroe image was to assume a qualitatively new style of expression. The studio publicists started to exploit the early material: a still was blown up nearly fifty feet tall and hoisted on to the front of Loew's State Theatre in New York. The director, Billy Wilder, began shooting the 'skirt' scene at 1.00 a.m. on a hot evening in Manhattan; the following report appeared in the *New York Herald Tribune:*

> Some 1,000 devotees of Marilyn Monroe watched carefully as Miss Monroe stood atop a Lexington Avenue subway grating and had her skirt blown high into the air fifteen times. One of her spectators was Joe DiMaggio . . . He had no comment on his wife's performance . . . However, other spectators whistled and cheered each time a huge portable blower situated under the grating sent Miss Monroe's white skirt skyward. Miss Monroe's garb included a white backless dress, white shoes, a white purse and white panties. She wore no stockings. Miss Monroe departed from the scene, in front of the Trans-Lux Theater at 51st Street at 4.15 a.m. after about two hours of work.

Not only did the experience sour Monroe's relationship with her image-makers, it also sparked off an argument with DiMaggio which attracted extraordinary press coverage. As the 'skirt' scene was shot, the crowd cried 'Higher! Higher!' As Monroe's dress swirled upwards her husband walked away. 'I'd have been upset,' said Billy Wilder,' if there were 20,000 people watching my wife's skirt blow over her head.' It seems that DiMaggio expressed how upset he was by physically harming his wife (see chapter 2). One of Monroe's publicists later said: 'Joe wasn't any great hero in Marilyn's life. He was vicious.'

The marriage break-up of Monroe and DiMaggio marked the tearing-up of a publicist's 'dream ticket': as the Los Angeles *Herald & Express* put it, the couple were the 'girl' who came to fame 'in a birthday suit' and the man who found success 'in a baseball suit'. Reporters called the couple, 'Mr and Mrs America'. By following the consequent divorce proceedings from contemporary press coverage, one can acquire a fascinating insight into the circuit of celebrity intrigue. The New York *Journal American* launched the 'story': 'Joe Fanned on Jealousy'. The press chief at Fox, 'fast-talking' Harry Brand, had extracted a promise from the newly married Monroe that, should the marriage ever end, she would tell him immediately. She is said to have complied. Brand mobilized his assistants: five employees (Roy Craft, Frank Neill, Chuck Panama, Mollie Merrick and Ray Metzler) were instructed to prepare a short press release saying that the break-up was the result of the 'conflicting demands of their careers'. Then Brand's staff telephoned four Los Angeles daily newspapers, the wire service, and such leading columnists as Walter Winchell, Hedda Hopper, and Louella Parsons, so that each 'would be the first to know'. 'Almost one hundred' reporters raced out to the DiMaggio residence in Beverly Hills. The journalists scrambled over the lawn, trampled down rose bushes, and broke branches from trees in order to establish unobstructed views for their cameras, and a crowd of onlookers gathered along the street. New York magazine *Tempo* said that if Brand could not come up with Monroe herself, it would be prepared to interview her stand-in. The Reno Chamber of Commerce sent a request for Monroe to obtain her divorce there. Monroe's lawyer, Jerry Giesler, emerged from the house and spoke to the press: they quoted him as saying that Monroe was upstairs sick in bed 'with a virus' while DiMaggio 'brewed a pot of soup for his ailing wife.' When a reporter asked why DiMaggio did not move out of the house, Giesler replied that he 'wouldn't be surprised if Joe stayed until the lease ran out.' The press began a further flurry of activity.

Columnist Sidney Skolsky, an old Monroe confidant, succeeded in reaching her for an 'exclusive report'. He quoted her as saying, 'There is no other man.' The other reporters, deprived of access to either party, interviewed the friends, interviewed the enemies, they even interviewed each other. Headlines emerged such as 'NIGHTS WORE DULL AT JOE AND MARILYN'S.' Newspapers encouraged readers to send in their opinions on the matter, and used the story to draw some tenuous links with 'general social issues' and 'modern marriage'. Louella Parsons had written of 'Joe and Marilyn' calling each other by 'smooch names'; she now 'revealed' that she had always known the couple were 'ill-fated'. Aline Mosby of United Press wrote that Monroe's recipe for happiness had been to 'serve Joe dinner in his chair while he watched TV', and let him 'wear the pants' in the family. Monroe, Mosby added, also acquired a 'king-size bed', because she disapproved of separate bedrooms and 'often in bed you think of something you want to say, and you're not going to chase down the hall to another room.' Self-styled 'sexy' gossip columnist Sheilah Graham was quick to match her rivals' revelations: 'Both parties were "bored right to the ears" with each other. . . Marilyn confided to friends: "Joe's idea of a good time is to stay home night after night looking at television." [He] objected heatedly to the fanfare of sexy photos.' *The Seven Year Itch* incident, reporters agreed, had 'caused' the divorce decision.

Days passed with Monroe and DiMaggio completely out of reach. Lawyer Giesler finally announced that Monroe would hold a 'silent' press conference: she would pose for pictures but would not speak to reporters. Whilst the press awaited her appearance, DiMaggio emerged and packed his luggage into his car, mumbling that he was going 'home' to San Francisco, and drove off down the road. Shortly afterward, Monroe walked out in a 'black, form-fitting dress'. The photographers surrounded her, snapped and shot her image, jostling for position, resembling 'a mob scene like something from the French Revolution'. Monroe burst into tears and was hurried away in a car by her lawyer. The Associated Press judged it 'an exit worthy of an Academy Award'. The pursuit began in earnest, with journalists visiting every possible sanctuary. The studio was suddenly swarming with 'investigative journalists'. The ubiquitous press agent Harry Brand was contacted, and his 'reply' was ready-prepared: 'We're all sorry at the studio that it happened,' he began. 'It was a wonderful kind of legend, Joe and Marilyn. Everybody loves 'em both. Everybody thinks it's Romeo and Juliet. It's the All-American Boy divorcing the All-American Girl' (in fact, Monroe divorced DiMaggio, not the other way round). A cynical Hollywood scribe inquired: 'But who gets custody of the Wheaties?'

The following day, a kind of 'normality' was restored. The press wound down its operations. Monroe returned to the set of *The Seven Year Itch,* clothed in puce pink pyjamas, going through 'one of the funniest scenes in the movie' with Tom Ewell. She looked leaner and paler when the rushes were viewed, but continuity was maintained. Despite her heartbreak, said a studio press agent, 'the show must go on.' 'Why?' asked an unusually astute reporter. The press agent replied: 'We're $50,000 and three days behind production on the picture already.' Monroe's movie obligations caused her during her career to suffer miscarriages and marriage breakdowns, and always to suffer in public. The 'DiMaggio affair' highlights the intensity of the media interest in Monroe's public and private life.

Uniquely, the stars of Monroe's rank were legally prevented from working on a freelance basis unless the studio consented, determining how and when and by whom the star's services were to be exploited (i.e. by the loan-out system). Because stars were constrained to offer services which centred on their presence rather than their performances, their circulation did not lead to a loss of competitive edge. Monroe, whoever she worked for, was always 'Marilyn'. The structuring of Monroe's performance on personality rather than character emphasized what was unique to her rather than what was generalizable to the craft—producing a performance that could only be impersonated, since it was too specific to be replicated. Since public interest has been channelled into stardom, attempts at impersonation only reinforced the 'uniqueness' of the star. Thus, the ideology of stardom, with its associated individualism, protected the studio from the risk of their property becoming public domain. Monroe also had a vested interest in being a 'personality': this emphasis excluded competitors and, in terms of contract negotiations, it allowed her to highlight the 'authentic' elements of the image on the screen. When Monroe came to form her own production company and renegotiate her relationship with Fox, this claim to authenticity proved crucial.

As a rational economic agent, Monroe was obliged to maintain as close a correlation between self, as appearance and personal style, and a 'viable' star personality image as was practicable. Whilst the contract for the employment of Monroe carefully stipulated the limits on the utilization of the 'Marilyn' image and the necessity for the preservation of its physical/psychological bearer as a matter of legal ownership, nonetheless it remained the case that Monroe was ultimately the possessor of the image, because it was linked to her very person. Thus, a legal monopoly confronted a *physical monopoly* in the bargaining relationship. The 'good' to be exchanged was the services of Marilyn Monroe for remuneration and the opportunity to further expose the stellar image. There existed a *de facto* situation whereby Monroe could seriously affect the costs of production by non-cooperation and other forms of temperamental 'sabotage' – a risk-laden strategy, but still a manoeuvrable space. Monroe, indeed, went on to exploit this space whenever she clashed with the studio.

Stars can circulate without necessarily losing their quality of being a distinct property. In the best of all possible worlds, a star's relationship to a narrative would be one of affirmation of a personality with box-office power. Marilyn Monroe was arguably the nearest approximation to this condition. As she became the *de facto* proprietor of her own image, she initially expected that the 'Marilyn' image would cease to project a simple personality and assume a reflexive (i.e. accountable) relationship to the actorly and communication aspects of her work. In fact, Monroe *did* begin to show a distance from her own image, which became an object for manipulation in relation to character (she joked about 'becoming Marilyn' for office-hours), and developed at the level of public interaction a self which was to some extent autonomous of star image and characterizations. The social meaning assigned to stardom under these conditions was that of professional communication in a position of 'representative' engagement with the 'real' world of the audience through the medium of the movies. Monroe's image, with her new independence, contained the import of social representativeness; she no longer simply engaged in a personality display whose commentary implications were a side effect constructed by the critics.

The emphasis on 'Marilyn', in a pragmatic system of manufacture, had the virtue of providing a concrete meaning to guide the process of production, exhibition and consumption. For these reasons, though the studios had trademarks for copyright purposes, emphasis in publicity was accorded either to the global evocation of 'Hollywood' or to stars such as Monroe. Handel (1950) points out that the American public did not see trademarks or genres as reliable means for judging films for the good reason that neither of these indicated that a particular quality of filmic experience was guaranteed, whereas stars, by continuity of image, did. Similarly, where stars failed to honour that guarantee, there was a tendency for unsuccessful films to be transcended by the continuity of the image (see Handel 1950; and chapter 2). Monroe was such a marketable star, in such a variety of media, that her attempts to change her image were seen by the studio as acts of willed destructiveness.

'Being a movie actress,' Monroe reflected, 'was never as much fun as dreaming of being one.' By the time Monroe came to make *The Misfits* at the beginning of the 1960s, she was caught in the contradictions of an industry keen to cut costs whilst anxious to protect its few remaining stars. Movies, now competing with television, could not afford to miss any opportunity for self-promotion. On location for *The Misfits* came the Magnum photographic team (led by Cartier-Bresson and Inge Morath), followed by Ernst Haas, Eve Arnold, Cornell Capa, Bruce Davidson, Elliot Erwitt, Eric Hartmann, and Dennis Stock. They were all contributing to a Magnum picture book, produced in conjunction with Viking's publication of Arthur Miller's screenplay. *Horizon, Harper's Bazaar, Life, Look* and *Esquire* negotiated pictures and articles. When production began, over two hundred people were scheduled for the location. Monroe, by this time experiencing considerable physical sickness and on the brink of separating from her husband, was hounded around the set by the more malicious of the press reporters. Liza Wilson, Hollywood correspondent for the *American Weekly*, sent some particularly cynical questions to Monroe. The queries included: 'Do you believe it is true . . . that it is very difficult, if not impossible, for a husband to be happy with an actress for a wife, because they . . . have so little time and attention to give their husbands?' Warming to her theme, Wilson then asked: 'I know that you have been very interested in psychoanalysis and have read many books on the subject. What do you think of the often expressed theory that many actresses are narcissistic and sexually frigid?' (In Goode 1986: 269) Monroe sent back the questions unanswered. She tried to organize interviews on her own initiative, yet even these met with editorial interference. Her fight to be seen as a serious performer was largely opposed by her employers and reporters. Indeed, Clifton Webb's insistence that Monroe was a serious artist (*Picturegoer*, 11 June 1955) received the response from his interviewer: 'All I can say is that I hope that this personality doesn't emerge on the screen.'

In Monroe, the contradictions implicit in being looked at are more starkly exposed. Apart from its public attrition, there is a public cost in any life made up of public performances: as Mailer (1973: 49) says, 'it is not a lack of grace that offers sexual problems when actors make love but the lack of an identity to give up to the act.' To be the focus of intense professional skill, to be looked at and lusted after, to be made beautiful and public, may have helped Monroe calm her misgivings about her body and its impermanence. However, the risk was never far away, either that she would disgust herself, or that she would discover disgust in the person who was supposed to be turning her into art – the cameraman, the director, the studio executive. Professional role-playing extracted from the performer some sense of her private identity; she spent much of her life making her self into a spectacle, seeing others make her self into *their* spectacle. It was surely this ambiguous feeling within Monroe that contributed to her quite profound alienation from her colleagues and her public.

During 1961, a series of telephone calls from an anonymous woman kept coming through to Tom Clay, a disc jockey on Los Angeles radio station KDAY. He eventually persuaded her to reveal her name. 'I'm Marilyn Monroe.' 'Oh, okay,' he replied, 'and I'm Frank Sinatra.' He put the receiver down. She called back several times, and he finally relented and agreed to visit the address she gave him. At 9.30 one morning he found the door opened by Marilyn Monroe, wearing a bathrobe and drinking champagne, looking extremely sad and disorientated. She told him that she spent the nights suffering from insomnia, and his radio shows were a form of company. For several weeks Clay became a sympathetic listener, telling her of his wife and young children. On one occasion which Clay notes as particularly memorable, he asked her: 'How can you be so lonely?' Monroe replied: 'Have you ever been in a house with forty rooms? Well, multiply my loneliness by forty. That's how lonely I am' (*Say Goodbye to the President*, BBC1 TV 1985). This alienation dominated her final few years. Her favourite young fan, James Haspiel (whose portrait was found alongside pictures of DiMaggio's and Miller's children after her death), recalls a late meeting with his idol in 1961. She was lonely, divorced, recently ill, and abandoned by her studio supporters. Unexpectedly, she produced a photograph for Haspiel, inscribed:

'For the one and only Jimmy, my friend. Love you, Marilyn.' Haspiel, by
now a grown man, was saddened by the implication that she had so few
friends.

The star is that level of meaning that supports the claim to give the
public what it wants without establishing an unrealizable identification of
such a claim with any given studio. This focus on stars and conflict over
stardom provides the gratuitous benefit of supplying a level of conflict
concordant with reconstructing monopolistic practices and corporate
power in individualistic competitive terms. It is, in other words, good
public relations. The permanent loss of a particular star was, in the
short-term, a tremendous blow to a studio's investment policy. Monroe's
death in mid-career provided her fans with a keener, more lingering sense
of loss than might be experienced for a star who died at the end of her
career. Equally, the death of the young and the gifted, from suicide or
from murder, is a familiar route to 'immortality'. The public sadness at
Monroe's passing attracted a 'second circuit' of celebrity–fan correspondence: May Ross watched a television tribute to Monroe by Dame
Sybil Thorndyke, and wrote to thank her for the gesture. Thorndyke
responded with her own note (12 August 1962), which the owner permits
me to quote from here:

> Thank you for your very kind letter. I think many people loved
> Marilyn simply for seeing her on the screen and she was a dear warm
> hearted girl and I think she is unique – we won't see her like again. It's
> good to see such faithfulness as yours to an artiste you have admired.
> Poor Marilyn.

This sense of waste and tragic vulnerability has informed many people's
perception of Monroe: a young woman fan explains,

> What strikes me now about Marilyn is how little she hurt people,
> given how unhappy, screwed up, disorganized she was. She really
> didn't seem to have a mean bone in her body, which is not to say she
> couldn't be selfish and drive people crazy and all, but that's different.
> (In Oppenheimer 1981: 93)

A related observation, with a sociological moral attached, was made by
Jean Cocteau when he said: 'This atrocious death will be a terrible lesson
for those whose principal occupation consists in spying on and torment-
ing the film stars' (in Taylor 1984).

Monroe's journalist friend, W.J. Weatherby (1976: 137), remembers
that

It was a common experience in New York to see someone resembling her. From the back, it looked like her image, even to the sexy walk. But when I crossed the street to take a closer look, a strange face always regarded me from under the blonde hair. She was an imitation Marilyn, one of many who saw their ideal self in her.

The *Daily Mail* (17 February 1979) featured an article entitled, 'I Know that Face', which went on to introduce 'The Girls Who Like Looking Like The Stars'. The reporter exclaims: 'Surely it's . . . no, it can't be . . . is it Marilyn Monroe? . . . Young Marilyn . . . is 20-year-old Elaine Heath from Beckenham, Kent. "I sell jewellery but I'd love to be an actress", she said.' The American performer, Peter Stack, began impersonating Monroe for friends at parties before developing his act into a professional revue. Even in London's Oxford Street during 1986, a Monroe lookalike was hired to promote a new fashion store: 'People could not believe their eyes. Even the traffic stopped' (*Today,* 31 October 1986). The resemblance was rhetorical: the image was part of a persuasive process which involved a producer, a personality, and a public.

There existed a kind of community of judgement concerning 'Marilyn', allowing her repertoire of affects to be circulated so very effectively. Her singular impact was formed by a source of emission, a channel of transmission, a point of reception. The communicative circuit continued even after the subject's death. Souvenir picture books collect Monroe's life as a series of fetishized memories, reproducing the same litany of 'facts' that convention deems to be the history of the figure. Confirming these texts are the stills themselves, privileged objects for our rapturous nostalgia. These vary as little from book to book as do the biographical summaries: a figure in white with the whirling clothes; a figure framed in gold, eyes shut, leaning forward, pouting; a naked body washed up on a beach of red velvet; a sad figure, hair shorter, a pained expression, eyes looking away. The 'Marilyn' image was a celebrity signature, a star profile for a public to purchase and preserve. Yet a semblance of inaccessible depth remains, a sense that here in this very public figure lies a profoundly private and personal quality, an authentic 'self' on which the star was styled. The tension between the person and the image remains a disturbing one, forever insinuating the feelings expressed by Paul Mayersberg: 'As a servant, the public adored her to the point of neglect. They built a pedastal for their mistress so high that they could not reach her, nor she them. The result was that she died of affection by adoration. To summarise simply: she did a lot for us, and we did nothing for her.' (In Taylor 1984)

'I don't think I've ever met a writer I'd like as my judge.'
Reproduced by kind permission of Sam Shaw, Camera Press (Lawrence Schiller), London.

6

Miller, Misfits and Mailer

*Coming out of the movie theatre, alone, mulling over my 'problem',
my lover's problem which the film has been unable to make me forget,
I utter this strange cry: not: make it stop! but: I want to understand
(what is happening to me)!*

Roland Barthes

*Those big tough guys are so sick. They aren't even all that tough!
They're afraid of kindness and gentleness and beauty. They always
want to kill something to prove themselves!*

Marilyn Monroe

A modest brownstone building in Brooklyn at the end of the forties: two
young, ambitious authors are fighting off fatigue with the prospect of
future fame. Both men are Jewish, both are driven by an innate belief in
their own abilities. One is writing a play which will eventually be called
The Death of a Salesman. The other is writing a novel to be entitled *The
Naked and the Dead*. They meet each other on the stairs occasionally,
'each certainly convinced that the other's modest personality would never
amount to much' (Mailer 1973: 19). One writer is Arthur Miller, the other
is Norman Mailer. Both men went on to find their fame: Miller was hailed
as 'America's greatest living playwright', whilst Mailer was proclaimed as
'the most important writer in America'. Both men found Marilyn Monroe
an 'obscure object of desire'. Miller married her, Mailer marvelled at her,
and both men mourned her. Their writings on Monroe have a wider
significance as dedications to the 'masculine myth' which works to secure
a meaning for Woman, to represent women as passive objects to be
looked at and longed for.

One of Monroe's New York friends was the poet Norman Rosten, a mutual acquaintance of both Miller and Mailer. She once said to him of her admirers: 'It scares me. All those people I don't know, sometimes they're so emotional. I mean if they love you that much without knowing you, they can also hate you in the same way' (Rosten 1980: 15). Rosten soon witnessed the disturbing pursuit of Monroe by her followers:

The crowd ... moves around her, moves away and closes in behind
her as she passes, a tide of living men and women who smell out the
one they will later crucify. They invariably pick a magical talisman:
contradictory, all-embracing, wife–mother–mistress, the mysterious link
between sex and death. For as certain as they adore her, they could
also destroy her. (p. 24)

Arthur Miller was not a fan but became an admirer and eventually married Monroe: he promised to care for her and admire her as others had not done, he proposed to respect her as an artist with genuine intelligence and skills. His relationship with Monroe revealed itself in his own work: she appeared, rewritten by him, in several stories, stories which showed the author striving to capture and comprehend the qualities in Monroe which had made him love her. The marriage eventually collapsed, and after Monroe's death Miller struggled to come to terms with an extraordinary repertory of sorrows, anger, frustration and guilt. Norman Mailer entered into this maelstrom, soon becoming absorbed in the subject and envious of Miller's very special involvement in the story. The two men struggled to claim 'Marilyn' as their own, their own personal property, and they scratched and scribbled over her memory with pens and ink and impassioned inscriptions.

The changing representation of woman in text and image circles around the unanswered question, 'What is she?'. Men typically appear as themselves, as individuals, but women attest the identity and value of someone or something else, 'other', and the beholder's reaction is necessary to complete their meaning. Meanings of all kinds flow through the figures of women, and they frequently fail to include who she herself is. Many men wrote 'Marilyn' and were consequently surprised when Monroe did not match this figure. Monroe was often made to speak through the words of men – including her husband. Arthur Miller would inform the press that his wife was certainly not a 'dumb blonde'; he would write a movie which could feature her demonstrating that she was no 'dumb blonde'. Yet rarely was Monroe afforded the time, the opportunity, or the respect in order to carve out her own personality and her own views; she suffered in silence, an unhappy silence described thus by

Cixous (1984): 'Delicacy of the silence pregnant with what one could say. Because happiness is not to say it: it is to be able to say.' It was a constant complaint by Monroe that she was unable to realise her own felt needs. Monroe was to the male writer 'a confirmation of something on the printed page to which one returns, fortified and satisfied' (Virginia Woolf, 1955). As she watched men become infatuated with their own image of her, she was moved to a kind of melancholy. Miller wrote of her as his 'child', his 'darling girl'. Mailer hailed her as the 'sweet angel of sex'. A contemporary advertisement for sanitary towels proclaimed: 'Film stars just can't have "off days". . .' As de Beauvoir (1984: 380) has said: 'It is not easy to play the idol, the fairy, the faraway princess, when one feels a bloody cloth between one's legs; and more generally, when one is conscious of the primitive misery of being a body.'

Monroe's romance with Arthur Miller seemed at first to offer her liberation from her studio role as a sex symbol. When they met she was a successful movie star, itemized as a unique individual. The studio reinforced her image as the woman in isolation, either in the posed pin-up or else as the epicentre of a contrived scene for a movie still. Very rarely was Monroe represented by an image easily recognizable by other women or by herself. She was consistently presented as either immacuately groomed and glamorous, or as the very opposite, dishevelled and unhappy to the edge of distraction. Her life, that which occupied her when she was not before the cameras or the press, was not seen. To isolate a person in this way is to render her powerless and to refuse any kind of perception of her as a member of society. Monroe was manoeuvred into a cocoon of self-interest and self-consciousness, the self as celebrity. When she made public her desire to study, she aroused feelings of betrayal in her employers. That many of her academic advisors found her to be an intelligent and witty woman was no consolation to the Hollywood men who found her ambitious and threatening. McCarthyism was sweeping through the industry: intellectuals were under suspicion of having left-wing sympathies. Arthur Miller was a well known 'leftist dramatist'. Hollywood was shocked and shaken by Monroe's alliance with him during the mid-fififties.

Miller was summoned to testify before the House on Un-American Activities Committee, and he was asked what was the objective of his planned visit to England in 1956. He said: 'My objective is double. I wish to attend a production of my play, and to be with the woman who will then be my wife.' Monroe was immediately descended upon by the press. She made a telephone call to Norman Rosten: 'Have you heard? He announced it before the whole world he was marrying Marilyn Monroe. Me! Can you believe it? You know he never really asked me' (Rosten

1980: 31). This was the first piece of masculine *hubris* performed by Miller, and it plunged Monroe's career into doubt; Simone Signoret (1979: 323) explained, 'One of two things could happen: her total destruction, or the rehabilitation in the eyes of the public of a man who, among others, was deprived of his passport; a man whose works were neither played nor published.' Monroe's presence by Miller's side in Washington during this period had a considerable impact upon public opinion. The affection in which she was held by Americans probably acted as a protective umbrella over Miller.

The marriage was profoundly precious to the movie star. Monroe had received tuition from Rabbi Robert Goldberg in order to formally become a Jew. At a ceremony she lifted her veil to sip from a goblet of wine, spoke her 'I do', exchanged rings with Miller, and they embraced each other. On the back of their wedding portrait Monroe wrote three words: 'Hope, Hope, Hope'. The newly married couple determined to spend much of each year at 'Arthur's farm' in Connecticut. In his study of the 'comedy of remarriage', Stanley Cavell (1981) has noted the prevalence of the 'green world', a place where the perspective of cultured people can be regained, psychic renewal obtained. This was always identified in Hollywood movies as 'Connecticut'. Miller, despite his Brooklyn background, was genuinely drawn to the land; for Monroe, it was an experiment, something more than a new role, but she was not sure that it would work out. When the country home was being renovated, they rented a cottage in Amagansett, Long Island, where Monroe threw herself into the task of making her marriage a success.

It was at Amagansett that Monroe learned she was pregnant. By the sixth week, she was in acute pain and Miller rushed her to the city, where a doctor diagnosed a tubular pregnancy and operated at once to save her life. She awoke from surgery to be informed that the embryo had been surgically removed. Doctors explained to her that the child she had lost would have been a son. She would never entirely recover from this experience and, ironically, she would die five years hence on the anniversary of this miscarriage. The desire to have children was intense for Monroe: her body would at last be her own, since she felt it would exist for the child who belonged to her, a physical exclamation of blood, spirit and soft, warm flesh. As a mother she hoped to have her existence justified by the wants she would supply: she would attend not a man but a self lost in a fragile and dependent body. Monroe's curiosity turned increasingly to her own body, her pride in it, and the urge to win it back for herself; the desire for a child became more acute, promising her commitment and continuity. She said to Truman Capote: 'I hate funerals. I'm glad I won't have to go to my own. Only, I don't want a funeral – just my ashes cast on waves by one of my kids, if I ever have any' (Capote 1981: 211). Arthur Miller remarked: 'To understand Marilyn best, you have to see her around children. They love her; her whole approach to life has their kind of simplicity and directness.'

That Monroe could ever have successfully carried a child is extremely doubtful, as she was plagued by continual menstrual and gynaecological problems. Nonetheless, her empathy with children meant that they gravitated to her unerringly, and she established affectionate relationships with both DiMaggio's and Miller's children by previous marriages. She could never truly stop herself from longing for a child of her own, yet this very need led to further suffering. As one of her friends put it, 'Her whole womb was weeping': whilst she was presented as an image of playful purity, inside she was bruised and broken, increasingly forced to use drugs to assuage her agony. Such was the pain that doctors tried to persuade her to consider a hysterectomy. She flatly refused, explaining: 'I can't do that. I want to have a child. I'm going to have a son.' Her stepchildren brought her some comfort and consolation; she responded to them with openness and real respect, telling them: 'See if I'm worth being a friend. That's up to you, and you figure it out after a while' (Meryman 1962).

Miller soon discovered how profoundly committed was Monroe's concern for life. Early in their marriage, walking along the beach in Amagansett, she saw some fish stranded on the shore; she began throwing them back into the sea, saying 'Some of them might live now till they're old ... and see their children grow up.' Miller wrote the short story, 'Please Don't Kill Anything' based on this incident. He had witnessed Monroe's sensitivity to death and her identification with animals dying because of man. He believed that this attitude was originally influenced by her childhood experience of being nearly suffocated by her mentally disturbed grandmother. Monroe felt with the intensity of a radio receiver turned to maximum volume: apparently distant, faint waves were amplified to the listeners' benefit, but the waves most local and strong threatened to overload the system. Monroe realized that while her fears may have seemed foolish to many, she could no more prevent these anxieties than fail to continue breathing. Montgomery Clift was astute in seeing Monroe as 'lacking a second skin': the unusual vulnerability attracted people at the same time as it gnawed at her existence. As Mailer was to write, 'when the wings of insanity beat thus near, one pays attention to a feather.'

Monroe suffered a miscarriage during the filming of *Some Like It Hot* in 1958. When pregnant she wrote: 'I'd love my child to death ... Arthur says he wants it, but he's losing his enthusiasm.' The physical pressures involved in making the movie made a major contribution to Monroe's miscarriage. She kept a photograph taken of her just before she had lost her baby; she said to friends, 'It was the happiest point of my entire life.' Speaking to the press, she commented: 'A man and a woman need something of their own. A baby makes a marriage perfect.' She succeeded to some extent in winning the affection of her step-children, but they were still unsettled by the taunts of their schoolfriends. As Steinem (1986: 85) notes, this side to Monroe was very important: calling herself Mrs Miller, Monroe struck up a friendship with a young Israeli woman, Dalia Leeds, who had a newborn son. The woman recalls that Monroe

> talked mostly about children. She was very curious about being
> pregnant, about what you fed a child, how you diapered it –
> everything. . . We laughed about having six children. She never
> confided in me about what difficulties she was having, but she very
> much wanted to have a child. . . She would play with the kids, hold
> them in her lap, and they adored having her.

In later years she would sit near the schoolyard and watch children play, painfully aware of her own solitude.

In the early period of Monroe's move to New York, she had made full acquaintance with her own powers, unconscious yet of anything but felicity and freshness in their exercise; and the kindliest, most compassionate side of her nature had found growth and encouragement. Her trip with Miller to England seemed to be a further move toward self-respect and professional prestige: working with Laurence Olivier on *The Prince and the Showgirl*, celebrating her marriage to Arthur Miller, she felt extremely optimistic. In fact, the visit was an unfortunate experience for all concerned.

Monroe and Miller took up residence in Englefield Green, Surrey. Miller was immediately drawn into a world of daily, hourly, crises and silent antagonisms, endless decisions and unpleasant arguments. With these tensions – which Monroe had been experiencing for several years as a major movie star – came the necessity of providing her with an almost constant support. It was an onerous task, a role which Miller soon felt ill-equipped to fulfil. Early in the making of *The Prince and the Showgirl*, Monroe came across some of Miller's notes lying on a table; they included comments concerning herself, and she found them deeply hurtful. She told her drama coaches, Paula and Lee Strasberg:

> It was something about how disappointed he was in me. How he
> thought I was some kind of angel but now he guessed he was wrong.
> That his first wife had let him down, but I had done something worse.

Olivier was beginning to think I was a troublesome bitch and that he, Arthur, no longer had a decent answer to that one. (In Guiles 1985: 227)

Miller had been hurt and offended when Monroe turned her anger and frustration onto him; he suffered the kind of alienation explained by Virginia Woolf in *A Room of One's Own* (1981: 35–6):

> Women have served all these centuries as looking-glasses possessing the magic and delicious power of reflecting the figure of man at twice its natural size. . . [I]f she begins to tell the truth, the figure in the looking-glass shrinks; his fitness for life is diminished. How is he to go on giving judgement . . . unless he can see himself at breakfast and at dinner at least twice the size he really is?

A year before marrying Arthur Miller, Monroe had been asked how she would define love. She replied that love was trust, that to love someone she needed to trust that person completely. Trust broke down during her honeymoon, and it was only partially restored when the couple returned to the USA.

Monroe's ectopic pregnancy sent her into a series of self-recriminations: was it her fault or was it fate? There was 'something wrong with her', something wrong inside her, a defect, an error, an evil which no promises nor prayers nor pills could correct. The sense of failure cut deep into Monroe's consciousness, and Miller was unable to help her overcome it. She was simply desperate to have children, she yearned to be a mother, she was eager to find fulfilment. Miller was also undergoing 'considerable anxiety', working his way through a writer's block. He eventually managed to complete some short stories; they included 'The Misfits', which was to grow into a screenplay for the final film Monroe would ever make. By 1960, when the movie was set to be made, Miller's relationship with Monroe had changed a great deal: he began to regard their marriage as in some way doomed. The movie's director, John Huston, had told Miller a story about a couple who separated after each told the other of their infidelities. Miller nodded, replying: 'Truth destroyed them.'

The Misfits was Miller's attempt at writing an elegiac tribute to Monroe's 'humanity'. In order to 'capture' her truth he had little hesitation in drawing on things she had told him in the privacy of their home. Miller said that Monroe 'lived on the front-line of life', and his own writing certainly encouraged her to do so. Rosten observed that the movie 'touched upon a raw nerve, a grinding connection between the fictional and the real life that exhausted her beyond her capacity to recover'. He added that Miller's 'disillusion, his bursts of tenderness, show in the film. They reveal his real attitude toward the marriage. And Marilyn did not flinch in her performance' (Rosten 1980: 78). During a scene in which divorce is discussed, Monroe's character is made to say: 'Husbands and wives are killing each other.' It is a line written by Miller and delivered by Monroe with an intensity that vibrated through the narrative, a confessional comprehended by both.

Miller's original short story, written in 1957, was Hemingway-like in the concentration on a man's world, causing *Esquire* to caption it, 'The Last Frontier of the Quixotic Cowboy'. What sound at the beginning like a celebration of sturdy, surly masculine independence becomes by the end an indictment of the times and even of the men themselves, misfits whose reiterated conviction that 'it's better than wages' partially obscures for them the sterility of their existence. A change of emphasis occurs in the movie version, effected by the introduction of a woman who had existed in the story only in the conversations of the men. Roslyn Taber (Monroe) divorces her husband in Reno, Nevada. Through her landlady, Isabelle Steers (Thelma Ritter), she meets Guido (Eli Wallach), an ex-mechanic who has been a lost man since his wife's death. Guido is attracted to her, but when he introduces her to cowboy Gay Langland (Clark Gable), the two fall in love. Gay is a strong individualist who plans to seek out and capture a herd of wild horses. He joins forces with Guido and Perce Howland (Montgomery Clift), an insecure young rodeo rider, and they agree to round up the 'misfits'. When Roslyn, now Gay's companion, learns that the horses are to be slaughtered and converted into dog food, she pleads with Gay to release them. He flatly refuses. Eventually, Perce responds to her request and sets them free. Gay, furious at this betrayal, recaptures the leader of the horses after a fierce struggle. Having demonstrated his power, he sets the animal free. Financially the round-up is a failure and the group split up, but each man has learned something from the ordeal. Gay and Roslyn, who have come to understand each other through the experience, agree to try to make a new start together.

Arthur Miller and John Huston seem to have followed Hitchcock's dictum: 'Torture the heroine.' Throughout her movie career, Monroe personified both the tragic *(The Asphalt Jungle, Niagara)* and comic *(Gentlemen Prefer Blondes, Some Like It Hot)* tensions of the sexual waif. By the time of *The Misfits,* when her increasingly turbulent personal life had become front-page news, there was a more acute sense of unease in her characterizations. In this movie, Miller juxtaposes rural America, the West, and the cowboy as embodied in Gable's character, to the neurotic frictions of city life represented by Monroe's Roslyn. Miller's script proved as psychologically shrill as his study of witch-burning and guilt

(The Crucible). There is a lack of humanity in his characters; the innocence and idealism of the thirties is displaced by a post-war aimlessness and lethargy. Though Miller sought to inject a scintilla of valour and romance in the ending, one realizes how little chance there is to sustain a movie marriage of Gable and Monroe, as was sadly emphasized by first Gable's and then Monroe's tragic deaths. Many critics were also convinced that Miller was working for a wish-fulfilment: he decided very late in the movie's making that Monroe's character would go off with the older man. Montgomery Clift was vocal in his belief that Miller identified with the figure of Gay. Monroe suspected Miller of trying to rewrite her own future.

The movie unit was riven by factions, bitterness and recriminations. The film's fatality seemed to outlast the production: days after the end of filming, Gable died from a heart attack; three of the other central figures, Montgomery Clift, Thelma Ritter and Monroe herself, were dead within a few years. The movie positioned Monroe in the most prominent and painful of places: her private comments were translated by Miller into his own screenplay, and his wife took on the aura of an optic poem, a 'text' to be read. Alice McIntyre (1961) watched her begin: 'she becomes at once the symbol of impartial and eternal availability, who yet remains forever pure – and a potentially terrible goddess whose instinct could also deal and whose smile, when she directs it clearly at you, is exquisitely, heartbreakingly sweet.' Miller had come to live with Monroe in the third person, as an observer. The more he wrote of her the further away she seemed. We seek to petrify evanescent aspects of our experience because we would otherwise have no hold upon them; it is only through the medium of the artfully wrought image that we can work back towards a discovery of that which, at such moments, we think and feel. Barthes argues that 'if you put the lover in a "love story", you thereby *reconcile* [her] with society. Why? Because telling stories is one of the activities coded by society, one of the great social constraints. Society tames the lover through the love story' (in 1976). (Barthes also remarked that 'fear is the misfit of every philosophy.') Miller had written a story for his wife: he was working to remember within words the woman he had fallen in love with, but he ended by losing sight of the person she now was.

Monroe performed whilst she was aware that her marriage to the man she had once admired and respected was on the point of collapse. She no longer truly respected him, no longer had faith in his abilities, in contrast to Clark Gable – who had figured in her and her mother's mind as the epitome of male beauty and dependability. She may, too, have been conscious of the fact that the character Miller had written for her had less strength, less reality than that of Gay. It was curious in this context that the more tormented Monroe's private and professional existence became, the more remarkable was her work and appearance on the screen. As Roslyn, she had undeniably become an actor of depth and subtlety and power which transcended a role which was not entirely realized in the writing.

Very soon into production, Monroe's screen role and her private life seemed to merge. There was so much of herself, her own psyche and sentiments, written into Roslyn that she found it increasingly difficult to put the role aside when filming stopped. She was also now heavily into pills – uppers (Benzedrine) and downers (Nembutal). It was not hard to understand why she wished to deaden the pain. Miller sometimes sat by her bed as she slept, anxious lest the drugs caused adverse reactions; eventually he would give up and move to another room. They spent less and less time together – Monroe retiring each evening after dinner, Miller working on the screenplay into the night. Montgomery Clift, homosexual and easily hurt, was drawn to identify with Monroe; she said of him that 'he's the only person I know who's in worse shape than I am.' Producer Frank Taylor observed that, 'Monty and Marilyn were psychic twins. They were on the same wavelength. They recognized disaster in each other's faces and giggled about it.' Monroe sometimes spent her rest days by sitting in a secluded restaurant above an old ghost mining town; the owner recalled, 'I often sit here and think about her. What a sad thing! Such a beautiful, lovely person. So sensitive.' Huston's script supervisor, Angela Allen, noted the strange effect Monroe was creating:

> She ate pills like children eating candy. She built up such tolerance for them, they didn't do much good, and then she would take more. . .
> What was so amazing about her was the thing she projected on the screen. She seemed so ordinary when doing a scene, and then you'd go to the rushes and see her up there, so different, like no one else. (In Guiles 1985: 284–5)

Roslyn is not so much a character as a quality. She has only the slightest of biographies (a former 'interpretative dancer' and occasional dance teacher), with no home, no husband, and no happiness. She has reached a level of vulnerability where she is no longer bound and love is no longer found, where all things carry the intimation of danger. Monroe fills the space left by the screenplay with a most poignant performance: she displays an exhilaration forever in suspense, trembling lest the next moment it may be wounded. Her special quality here is that the pleasure and elation are always held tremulous on the edge of a precipice of doubt, fear, and extraordinary vulnerability. Throughout her career, and culminating here, even when she was at her most sexually aggressive and triumphant, one never loses sight of the tender, fragile, frightened human being beneath the skin.

W. J. Weatherby (1976: 54) appreciated this quality when he visited the set: 'she worried her way into a scene, into re-creating an emotion. Take after take was needed until she was ragged enough to live the scene. It was an approach that was very hard on her, her director, and her fellow actors, and everyone looked for the early signs of another breakdown.' On the set of her previous film, a notebook was found in which she had written: 'What am I afraid of? Why am I so afraid? Do I think I can't act? I know I can act but I am afraid. I am afraid and I should not be and I must not be.' She did not have to turn her self inside out for the role of Roslyn: Miller had already done so in writing the character. Monroe was exposed as never before, and Montgomery Clift could sympathize with her experience: 'Marilyn was an incredible person to act with . . . the most marvellous I ever worked with. . . But she went over the fringe. Playing a scene with her, it was like an escalator. You'd do something and she'd catch it and it would go like that, just right up' (in Taylor 1984).

In none of her other screen performances is her character so delicately shaded, as finely tuned, as warm and compassionate, as full of feeling as in *The Misfits*. It seemed as though she appreciated that the emotions she was conveying to the camera would be received by audiences with true understanding and affection. Her actions appeal to the audience, her words connect to her eyes, and the unspoken plea is, 'Stay near me.' Natalie Wood observed: 'When you look at Marilyn on the screen, you don't want anything to happen to her. You really care that she should be all right' (in Taylor 1984).

Monroe's character in the movie has the abstraction, and the intimacy, of a figure and an object in a dream. Miller had once said of Monroe: 'Being with her people want not to die.' Roslyn is Monroe inflected by a confused but devoted lover: she is Miller's image, the tender, torn tissue from which the pulse of conscience comes. Reminiscent of Cordelia, Roslyn 'redeems nature from the general curse'. There are certain exchanges in the movie which impart vivid insights into her personality. Perce is struck by Roslyn's peculiar gentleness: 'How come you've got such trust in your eyes, like you were just born?' Gay experiences the melancholic aspect of her character: 'You're the saddest girl I ever saw.' She replies: 'You're the first man who ever said that. They usually tell me how happy I am.' The viewer sees the scene through Monroe's biography like one studies a distant figure through binoculars. Roslyn's horror, misery and her outburst at the inhamnity of the men to the mustangs is founded upon the profound and empathetic sensibility which Monroe had shown in respect of animals, and her playing of these scenes is electrifying. In lighter scenes, the intense joy in her physical being generated an elation which is an irresistible source of her charm. Walking up and down the solitary, makeshift step leading into the unfinished house is beguiling: home, delight in being loved, a path taken often and here taken with such freshness and hope, exuding a transcendent spirit. Gay looks at her and says: 'You have the gift of life. The rest of us are just looking for a place to hide.' Guido asks her: 'How do you get to know anyone kid?' The question carries elements of Miller's own obsession, an obsession blind to the quality of sympathy found in Benjamin's dictum (1979: 77): 'The only way of knowing a person is to love them without hope.'

Insecure, inarticulate, Roslyn is vitally incarnate. Dancing with Guido, she hears his wife had no gracefulness. Uncomprehendingly, she asks: 'Why didn't you *teach* her to be graceful?' When Guido questions the possibility of achieving such an aim, she argues: 'if you loved her you could have taught her anything.' The naive sincerity of this comment is the emotional equivalent of her bodily movements – simple, flowing and spontaneous. In a world where nature has been devalued, where work has been degraded, where, as Gay laments, 'it just got changed around, see?', Roslyn represents the few precious unchanging principles. In one scene she breaks away from the group and moves over the grass and embraces a tree: given over to occupations, friends, memories, she closes her eyes, listens to the blood pulsating in her ears, loses herself in some secret pleasure or pain, shutting herself within an anonymous life which subtends her personal role. This scene is soon over, but the impact is immense: locked within a passive body, enwrapped in a blissful lethargy, thus imposing a temporary respite upon the constant drive which projects the body into things and towards others. The men look on, unable to comprehend such sensuality.

Roslyn is depicted as 'the other' amidst the barrenness of the desert; the source of hope, the Woman, the space from which men originate. The common myths are drawn upon by Miller: he makes Roslyn appear to possess a secret, pure presence in her heart which reflects the truth of the world. 'Honey,' Gay says, 'when you smile it's like the sun comin' up.' She is not creative yet she fructifies. She heals and she nurses, she knows all about men that hurts their pride and harms their self-esteem, she softens the hard angles of men's constructions and bestows upon them unforeseen grace. She wafts the breath of poetry through the lives of men: she resembles a poetic reality, for men project on to her all that they resolve not to be. This woman is isolated in a world of men, an object for the gaze of men. Miller had called Monroe 'the most womanly woman', and in *The Misfits* he tried to inscribe this image. The result was a man's idea of femininity, one that made 'a man feel good', and one exposed by Adorno (1978: 95) amongst others:

> The feminine character, and the ideal of femininity on which it is modelled are products of masculine society. The image of undistorted nature arises only in distortion, as its opposite. Where it claims to be humane, masculine society imperiously breeds in woman its own corrective, and shows itself through this limitation, implacably the master. The feminine character is a negative imprint of domination.

Innocent, trusting, tender, unsure of herself but sure of the possibilities of life, in sympathy with the natural world, Roslyn can easily be seen as the idealized Woman of the masculine imagination. That she transcends this stereotype is mainly through the quality of Monroe's performance, but also because of the effect she has on the other characters. She is at one and the same time an attraction and a distraction, a disturbing, disruptive force as well as a soothing, regenerative one. That Roslyn persuades the men, at the end, to release the mustangs constitutes another change from the original situation; that in doing so she seems to be restoring the men's liberty as much as the mustangs' gives the ending a note of hope, intensified by the final suggestion of a lasting relationship between Roslyn and Gay.

Monroe criticized Miller for making the Roslyn character 'too passive'. Roslyn has a need, but no way of satisfying it; there are words, but nothing to work them on; there are hopes, but nothing to pin them on. She has the mystique of the Muse. Monroe was made to play a figure perceived and positioned by men, a frustrating role well-expressed by Kristeva (1977: 15): 'voice stilled, body mute, always foreign to the social order . . . voice without body, body without voice, silent anguish choking on the rhythms of words, without sounds, without images, outside time, outside knowledge.' Each man looks on Roslyn as the one who can change his life and make him whole. Each man wants her love, her care, her concern. She is slowly being torn apart by the needs of others. None of these men seems to care much for what *she* needs, wants or wishes. None of them consider what can make her 'whole'. Yet her needs are more desperate than theirs, as Monroe's were more desperate than Miller's. When the misfit men are tying up the last of the wild horses destined for dog meat, Roslyn/Monroe screams: 'You are the killers!' In her eyes these men are the only true misfits, the tragic misfits, without values or virtues. She, at least, cares for something other than herself.

Roslyn had even been distressed when Gay proposed shooting a rabbit that had nibbled some lettuces: 'Couldn't we wait another day and see? I can't stand to kill anything . . . it's alive and . . . it doesn't know any better, does it?' Roslyn knows her separateness: her poignance comes from her acceptance of the unacceptability of her love, and by nonetheless sustaining her love and the encompassing knowledge it brings. She believes nature not to be the cause of evil – there is no cause in nature which makes these hard hearts, and no cause outside either. The cause, one feels she senses, is the having and holding of a heart in a world made heartless. Roslyn, talking with Gay after an attempt to resolve their differences, whispers: *'Love me.'* Alone, immediately after, she softly says: 'Help'. Monroe had once written a poem which included the lines:

Help Help
Help I feel life coming closer
When all I want is to die.
(In Rosten 1980:5)

Her final major scene as Roslyn has her run away from us, deeper and deeper into the depths of the cinema screen, into memory, towards eternity. She screams out: 'You liars! All of you! *Liars!* Men! Big men! You're only living when you can watch something die! Kill everything, that's all you want! Why don't you just kill yourselves and be happy? I pity you!' The cry emanates from deep within her, through the throat, with little help from lips or tongue: a visceral response, a great refusal (truth is killing her). When one has followed Monroe's biography, her career, her broken marriages and her treatment at the hands of men, then one can fully appreciate what a *triumph* this final scene is.

In retrospect, *The Misfits* seems an uncommonly cruel piece of writing. It required Monroe to go before the camera and virtually flay alive her inner self. As she said to Goode (1986: 200): 'I can't memorize the words by themselves. I have to memorize the feeling.' The harsh desert location was equally unforgiving, and delay followed delay as the star of the piece struggled to deliver herself through a haze of barbiturates and terror. Miller had become involved during filming with Inge Morath, a photographer covering the movie for Magnum. Monroe had sensed something between her husband and the woman who was later to become his third wife. That Monroe was still able to assume her fictional mask during this period was an act of courage, a triumph of will and professionalism. The viewer of *The Misfits* can see Miller's fingerprints on Monroe's emotions; the *New York Herald Tribune* carried the comment,

There are lines one feels Miss Monroe must have said on her own. . .
In this era when sex and violence are so exploited that our sensibilities

are in danger of being dulled, here is a film in which both elements are as forceful as in life but never exploited for themselves. Here Miss Monroe is magic but not a living pin-up dangled in skin-tight satin before our eyes. . . And can anyone deny that in this film these performers are at their best? You forget they are performing and feel that they "are".

The death of Clark Gable soon after the end of filming was a shock for Monroe. She had developed a real friendship with him: 'The place was full of so-called men, but Clark was the one who brought a chair for me between the takes.' When reporters suggested that the difficulties she had caused on the set may have hastened his heart attack, she was deeply distressed and fell into a depression. A newspaper headline of the time described her as 'The Golden Goddess who cracked'. Arthur Miller had been involved with Monroe each and every day of their marriage, using up more emotional energy than many people use in a lifetime. He was exhausted, and he withdrew. Monroe went to Mexico where her marriage was legally terminated. Miller sat in a viewing theatre, stunned: 'I still don't understand it. We got through it. I made a present of this to her, and I left it without her.' Monroe commented: 'He could have written me anything, and he comes up with this. If that's what he thinks of me, well then, I'm not for him and he's not for me.' She added to reporters: 'He is a wonderful writer, a brilliant man. But I think he is a better writer than a husband' (in Taylor 1983: 52). Asked by a journalist to comment on the common belief that Miller had originally 'sought her out' because he had 'come to an end in his writing', Monroe paused. Before answering she insisted that her response be printed in its entirety. The reporter agreed, and she replied: 'No comment'. She told Norman Rosten that she had arranged with Miller for her to visit their house and collect some objects of sentimental value:

> I told him when I'd be there, but when I got there he wasn't. It was sad. I thought he'd be there and maybe ask me in for coffee or something. We spent some happy years in that house. But he was away. And then I thought, Maybe he's right, what's over is over, why torment yourself with hellos? Still, it would have been polite, sort of, don't you think, if he'd been there to greet me? Even a little smile would do. (Rosten 1980: 83)

'Things always work out for a while,' she said, 'and then. . .' When Monroe's first husband, Jim Dougherty, heard of Monroe's death, he turned to his current wife and whispered: 'Say a prayer for Norma Jeane. She's dead.' During the last interview she gave, she had expressed her anger and sadness over the behaviour of certain men, those 'big tough guys'. Among her possessions was found a letter, unsent, to DiMaggio: in it Monroe had written:

> Dear Joe,
>
> If I can only succeed in making you happy – I will have succeeded in the biggest and most difficult thing there is – that is to make *one* person completely happy. (In Guiles 1985: 376)

Her death was received with worldwide grief, leaving an indelible mark on everyone's image of Monroe. Benjamin (1979: 98) writes, on first hearing of someone's death:

> there is in the first mute shock a feeling of guilt, the indistinct reproach: did you really not know of this? Did not the dead person's name, the last time you uttered it, sound differently in your mouth? Do you not see in the flames a sign from yesterday evening, in a language you only now understand?

Arthur Miller said to the press: 'It had to happen. I didn't know when or how, but it was inevitable.' There is a hint of absolution in the idea of an 'inevitable' end: one is 'less responsible', less vulnerable to feelings of guilt. After a period of silence, Miller attempted to 'prove' his theory: the narrative was founded on inevitability as a 'determined' account. *After the Fall* saw Miller's own anguish dyed blonde, washed over by ill-conceived apologies.

According to Weatherby and others, Miller's new wife, Inge Morath, persuaded him to write the play in order to lay the ghost to rest and allow their own marriage to grow. The play appeared in 1964, directed by Elia Kazan, an old colleague of Miller's. Simone Signoret (1979: 324–5) wrote: 'it's sad that a Kazan–Miller reassociation was celebrated across a box called a coffin. A coffin for the blonde. It seems to me that they disfigured her, at least in part; in any case, they betrayed what was best in her.'

Shortly after the play opened, *Life International* (24 February 1964) announced on its cover 'Marilyn's Ghost. Arthur Miller writes about his shocking new play.' Inside the magazine, Tom Prideaux asked pointedly whether too much was revealed and concluded that Quentin, the 'Miller' figure, 'protests too much in the kind of moral strip-tease' that he performs. Miller's own article was entitled: 'With respect for her agony – but with love'. He expressed surprise at the public and critical outcry, which he saw as hypocritical; he rounded against those who charged him with 'cruelty toward the memory of Marilyn Monroe', and he denied that the 'Maggie' character was a representation of Monroe. According to Miller, Maggie was only 'a character in a play about the human animal's unwillingness or inability to discover in himself the seeds of his own destruction.'

However, as Prideaux observed, no matter how much Miller repeated that Maggie was not 'Marilyn', 'the play itself invites the comparison – not invites, insists on, really.' The story features a lawyer, Quentin, talking to the audience as his stream-of-consciousness conjures up figures from his past – most memorably, Maggie, his pop star wife who committed suicide. Kazan and Miller, according to Weatherby's testimony, went to great lengths to make the actor Barbara Loden resemble Monroe – a blonde wig, some familiar mannerisms, similar clothes. Weatherby notes (1976: 219): 'There are too many incidents and remarks that recall the Miller–Monroe relationship for the play not to seem a revelation of what went on behind the scenes and to be Miller's view of what went wrong.' It thus was quite disingenuous of Miller to express the hope that, within the near future, his play would be separated from 'Marilyn's golden image'. To admirers of Monroe it seemed that Miller, persecuted by the memory of his dead wife, was projecting his guilt-feelings on to her after-image, 'after the fall'. 'Marilyn' became an internalized grudging and envious object, disturbing his work, life and love. Despite Miller's claims to the contrary, there is a shocking absence of love in this play.

Disillusion, divorce, and Monroe's death must, inevitably, have raised acutely for Miller problems of personal responsibility over which a work of this nature could normally have drawn a veil. For one whose plays have always been rooted in a bloodied skein of guilt, however, the impulse to explore these issues dramatically was perhaps equally 'inevitable'. He dedicated the play: '*For my wife*, Ingeborg Morath'. At one point the Quentin character says: 'I am bewildered by the death of love. And my responsibility for it' (Miller 1979: 64). The play is concerned with Miller's suffering, not Monroe's. He appeared honestly to include some of his own destructive acts, such as the 'diary' episode from their honeymoon when Monroe discovered his notes about her. She wondered, was Miller exploiting her, as most men had done, but in a more 'intellectual' way? In *After the Fall* a similar scene occurs; Maggie angrily tells Quentin not to mistake her for his former wife, and he replies: 'That's just it. That I could have brought two women so different to the same accusation – it closed a circle for me. And I wanted to face the worst thing I could imagine – that I could not love. And I wrote it down, like a letter from hell.' (p. 114)

The fey, trusting innocence of Roslyn has given way to Maggie's egoistic and unfeeling arrogance. The rawness of Maggie/Marilyn's

agony is more apparent than the 'respect' and 'love' which Miller claimed
had prompted him to the depiction of it. The relationship between
Quentin and Maggie culminates with the realization: 'Maggie, we were
both born of many errors; a human being has to forgive himself! Neither
of us is innocent. What more do you want?' Fearful that he himself cannot
love, Quentin tries to force from Maggie the admission: '"And I am full of
hatred; I, Maggie, sweet lover of all life – I hate the world!"'

The 'Marilyn' image is for Miller the blessed woman of his dreams and
the cursed woman who belied those dreams. He projects on to Maggie
that which he desires and fears, his love and his hatred. He seeks the
whole of himself in her, and the fact of her otherness frustrates him. In this
image one sees the gap between an author's world view and his egoistical
dreams. Quentin is made to say: 'I have to survive too, honey.' 'A suicide
kills two people, Maggie, that's what it's for! So I'm removing myself, and
perhaps it will lose its point.' 'You want to die, Maggie, and I really don't
know how to prevent it.' This is tortuous writing: the persistence of
Monroe's image is a permanent wound in the memory of Arthur Miller.
He resorts in *After the Fall* to arguing that guilt is within us all, and there
follow several allusions to the Holocaust: 'Who can be innocent again on
this mountain of skulls' asks Quentin. No one can condemn Arthur Miller
for wishing to come to terms with the confused feelings of guilt and
resentment after the death of both a marriage and a lover. However, to
relate his individual experience to that of his race is an intolerable act. The
Holocaust, as Adorno (1973: 361) has shown, 'defies human imagination
as it distills a real hell from human evil'. It is unfortunate that Miller
should wish to use this horrible image in such a personal context. There is
a strong case for saying that there can be 'no poetry after Auschwitz'.
Monroe's extraordinary belief in the value of life as such was quite
astonishing when seen in this context. For Miller to use both her death
and the death of millions of 'specimens' for the working-out of some
literary problem is an act which borders on the offensive. One must
consider the unforgettable, heartrending comment by Elie Wiesel (1981:
126), made after 'surviving' Auschwitz and Buchenwald:

From the depths of the mirror, a corpse gazed back at me.

After the Fall reveals Miller struggling to continue writing 'after
Marilyn': the result is a regrettable semi-confession which seems to bury
Monroe rather than praise her. The fiction is an ugly one, gnarled by the
kind of guilt and recrimination experienced by Monroe at the end of their
marriage:

When we were first married, he saw me as so beautiful and innocent
among the Hollywood wolves that I tried to be like that... But when
the monster showed, Arthur couldn't believe it. I disappointed him
when that happened... I put Arthur through a lot, I know. But he also
put me through a lot. It's never one-sided. (In Weatherby 1976: 187)

The man who saw Miller as a bitter rival had been observing this
relationship from a distance. Norman Mailer had followed Arthur Miller
into the forefront of the New York cultural establishment. Norman
Rosten writes of Mailer's exasperation over his inability to arrange a
meeting with Monroe. When Mailer stayed briefly near Miller's farm,
Monroe was working elsewhere. When Mailer pressed Rosten to
organize a communication, the request was rejected by Monroe: she told
Rosten that 'one writer is enough', that she 'had nothing to say', and that
Mailer was 'too tough'. Discussing Mailer's novel of Hollywood life, *The
Deer Park*, Monroe told Weatherby that Mailer was 'too impressed by
power'. Predictably, Mailer never met Monroe; he says he spent several
years, waiting 'for the call to visit, which of course never came'.

Mailer spent the years of Monroe's movie stardom in a Hemingway
haze of dangerous delusions, a state which culminated in his assault on his
wife. Working through this guilt, the Faustian advocate of psycopathy
evolved into an apocalyptic pundit of the 1960s, a decade that embraced
him and his harangues. By the 1970s, having written *The Prisoner of Sex*
and now confronted by 'every man's love affair with America', Mailer's
misogynism was finally under analysis. His approach is a notable example
of the way in which many modern male authors define themselves as
artists through a sexual dialectic, create explicitly masculine modes of
speaking and writing, and contend with feminism and male and female
rivals. In the modern era, writing and the cultivated use of language have
sometimes been viewed as activities so close to the 'effeminate' that some
male writers have felt an urgent need to establish their virility through a
variety of strategies, often involving the denigration of women (see
Schwenger 1985). For Mailer and his imitators, the threat of feminization
led to a masculine mode of writing – tough, curt, slangy, obscene – which
could absolve itself from the self-conscious, and therefore unmanly,
refinement of art.

Mailer betokens an anxiety about writing as a feminine craft, in which
verbal dexterity and the desire to please risk effeminacy. Worried that his
first-person narrators were 'over-delicate', Mailer rewrote the first
version of *The Deer Park*, deciding 'which of two close words was more
female or more forward'. With each rejection of the 'female' word, he
noted, 'the style of the work lost its polish, became rough, and I can say
real, because there was an abrupt and muscular body back of the voice

now.' With *Marilyn*, Mailer's determination to inscribe his vision of a formless female sexuality is evident on almost every page, with nearly every vaginal metaphor: she is 'soft in her flesh' (p. 16); her kisses are 'like velvet' (p. 46), she is ready to 'drop your body down a chute of pillows and honey' (p. 102). There are few writers maler than Norman Mailer: as Kate Millet remarked, 'he seems to understand what's the matter with masculine arrogance, but he can't give it up.' His writing works to keep everything under surveillance, to see clearly, hence to dominate through vision. Mailer wrote two books on Monroe – one a biography, the other a pseudo-autobiography. In 1981, he even wrote a play about her, *Strawhead*. The attraction to 'Marilyn' became an imaginary affair of the heart and mind. *Marilyn* (1973) was Mailer's longest and most compelling meditation on celebrity, the introduction which grew into the book, the deepest caption in the history of picture books.

From the beginning, Mailer was perplexed at the prospect of loving someone without hope. He wrote (1973: 19) of Monroe's biographer, past and present: 'No matter how much he could learn about her, he could never have the simple invaluable knowledge of knowing that he liked her a little, or did not like her, and so could have a sense that they were working for the same god, or at odds.' He concluded that the most appropriate medium in which to meditate on Monroe was 'a *species* of novel ready to play by the rules of biography'. His aim was to produce (p. 20):

> a literary hypothesis of a *possible* Marilyn Monroe who might actually have lived and fit most of the facts available. . . A reasonable venture! It satisfied his fundamental idea that acquisition of knowledge for a literary man was best achieved in those imaginative acts of appropriation picked up by the disciplined exercise of one's skill.

Mailer's study results in the (unsubstantiated) claim that Monroe was murdered by a right-wing conspiracy to embarrass Robert Kennedy. In the light of subsequent findings, Mailer's intuitive work is impressive. However, his fascination with power and his weakness for the great Kennedy 'ladykillers' combine to mute his reaction to this image of a death. He seems a man closer to politics than to poetics; *Marilyn* contains more charisma than characters, or cogency, or charity, or even simple courtesy.

As he was writing the biography, Mailer found himself focusing on the figure of Arthur Miller – not so much as a person but as a threat to his own 'relationship' with Monroe. Miller, wrote Mailer, 'had only a workmanlike style, limited lyrical gifts, no capacity for intellectual shock, and only one major play to his credit'. Soon the pretence of good form was over (p. 143): 'find the witness to testify that Miller had ever picked up a check.' The other side to Mailer, the side which exhibits a welcome honesty, moved him to confess (pp. 19–20):

> The playwright and the novelist had never been close . . . the secret ambition, after all, had been to steal Marilyn; in all his [i.e. Mailer's] vanity he thought that no one was so well suited to bring out the best in her as himself, a conceit which fifty million other men may also have held. . . It was only a few marriages (which is to say a 'few failures) later that he could recognize how he would have done no better than Miller and probably have been damaged further in the process.

Mailer's appreciation of Monroe was a significant development for him, reminiscent in some ways of the transformation undergone by Gay in *The Misfits* when he confesses: 'You just shine in my eyes. . . I think you're the saddest girl I ever met.' Gay concludes by saying: 'I never bothered to battle a woman before. And it was peaceful, but a lot like huggin' the air. This time I thought I'd lay my hand on the air again – but it feels like I touched the whole world.'

Mailer finished the biography believing that he could identify with much that was integral to 'Marilyn'. He acted on this conceit and wrote a 'Marilyn autobiography': *Of Women and their Elegance* (1980). Explaining the text, Mailer wrote (p. 284): 'It arises from certain facts, and there are several sections within it that are all made up, and it cannot be said that the fact is wholly factual in the other places. . . Perhaps we may call this an imaginary memoir, an as-told-to book, a set of interviews that never took place between Marilyn Monroe and Norman Mailer.' Writing as Norman Jeane/Marilyn Mailer, he ensures that 'she' presents herself as a closed surface, excluding what can contradict its supposedly seamless feminine unity. This 'Marilyn' is too biographical to be pornographic: we are drawn to her haecceity. Nonetheless, the end result is a mythic version of Mailer's woman, still a 'sweet angel'.

Monroe's favourite expression in recounting experiences with men was that they were rarely 'there' – meaning, they rarely cared sufficiently to consider what she thought or needed beyond their own self-interests. Evident as Miller and Mailer are in their rewritings of 'Marilyn', they are very rarely 'there'. With a comment reminiscent of *The Misfits*, Monroe remarked: 'Lies, lies, lies, everything they've been saying about me is lies.' Imputation masquerades as interpretation: the men tell us what 'Marilyn' wants, needs, and feels. Making *The Misfits*, Monroe said of John Huston: 'He treats me like an idiot – "Honey, this" and "Honey that".'

These men refrain from allowing Monroe to speak for herself: they announce her problems and suggest solutions. Monroe is thus defined, depicted, displayed and designed *by* men: she is the object, they the creators, she is the commodity, they the consumers. Simply from the marketing point of view, a multiplicity of 'Marilyns' is an egregious attraction for the publishing houses.

Monroe became increasingly suspicious of men in her later years. She confided to her old photographer friend, André de Dienes: 'They've all exploited me and now I've got nothing' (de Dienes 1986: 154). Her personality became significantly melancholic – a feature frequently ignored or underplayed by her biographers. She was found in a despondent state of mind during 1960 by her maid, Lena Pepitone; Monroe cried, 'Lena, no. Let me die. I want to die. I deserve to die. What have I done with my life? Who do I have?' (Pepitone and Stadiem 1979: 165). Her work was frustrating her: 'I feel like I'm rejecting part of myself, that I'm letting part of me die, like a dead branch that gets no chance to grow and develop' (in Weatherby 1976: 188). Jack Lemmon was a fellow actor who sensed Monroe's melancholy:

> I think there was always a lot of facade when her mood seemed to be gay. Underneath, I think she had been very hurt and. . . I don't think she would let many people get close any more. When she liked someone, I think she'd let you get just so close and then a screen would drop because she didn't want to get hurt again. (In Freedland 1985: 61)

One man who was sympathetic to this aspect of Monroe's personality was Norman Rosten. She confided to him: 'I used to write poetry sometimes but usually I was depressed at those times. The few I showed it to . . . said that it depressed them – one of them cried but it was an old friend I'd known for a long time' (Rosten 1980: 14) She had been having treatment for a number of complaints at this time. The doctor who treated her for abnormal bleeding from the uterus and an ulcerated colon believed that the latter condition was a result of 'a chronic fear neurosis'; he said his patient was 'highly nervous, frightened, and afraid' (Cottrell 1965). Recuperating, she spent time in art galleries accompanied by Rosten. She was particularly fascinated by a Rodin statue depicting the figures of a man and a woman in an impassioned embrace. She kept shaking her head, saying: 'Look at them both. How beautiful. He's hurting her, but he wants to love her too' (in Rosten 1980: 104). Weatherby also heard Monroe's sadness; she explained: 'I was never used to being happy. For years I thought having a father and being married meant happiness. I've never had a father – you can't *buy* them! – but I've been married three times and haven't found permanent happiness yet' (Weatherby 1976: 147). *The Misfits* had ended with Monroe and Gable driving deeper into the night, following a star which would guide them home. Months later, recovering in hospital, Monroe gazed out at the night sky and murmured: 'look at the stars. They are all up there shining so brightly, but each one must be so very much alone.'

After Monroe's death, her final movie acquired a profoundly prophetic quality. Watching her in *The Misfits* is to touch the curve of helplessness one feels when one watches someone die. The role of Roslyn, we know with hindsight, is agony for Monroe, drawing on her innermost anxieties and insecurities, peeling off successive emotional skins. The camera is a 'clock for seeing with' (Barthes 1981), and in this instance we see each successive scene as a moving minute hand which hangs over the life on display, causing us to give it a fleeting, fearful glance. We are reading Monroe backwards, her past condensed into its spatial forms, its premonitory structures. She leaves us, presently, like the passenger in the moving train. What I reveal here is what I share with everyone else present with me at what is happening: that I am hidden and silent and stationary, that there is a point at which I am quite helpless before the acting and the suffering of others. What is painfully evident is my separateness from what is happening to Monroe: that I am I, and I am here. For all you and I can see, for all we know, the motion in this movie is consuming Monroe's existence from within ('I feel life coming closer'), drawing it out only to call it back in, and the ineluctable sensation as we view the last few moments of the movie is that we are witnessing the end of 'Marilyn', that *She is going to die*. We have witnessed something now that is no more. This is an enormous trauma for the audience: each viewing of *The Misfits* is implicitly, in a repressed manner, a contact with what has ceased to be, a contact with death. The experience is of a fascinating and funereal enigma. It touches the tenderest part of our relationship with Monroe – the knowledge that we can never help her.

During her final interview, Monroe had said: 'Everybody is always tugging at you.' By that time, in truth, they were not just tugging at her, they were tearing her apart. Montgomery Clift observed the danger he shared with Monroe:

> We attract people the way honey does bees, but they're generally the wrong kind of people. People, who want something from us, if only our energy. We need a period of being alone to become ourselves. To be an actor, you can't afford defenses, a thick skin. You've got to be open, and people can hurt you so easily. (In Weatherby 1976: 75)

There are moments in *The Misfits* when the plot seems to dissolve and we feel we are witnesses 'backstage'. We observe Monroe in the movie, somehow present to the proceedings – not spying, almost as if dreaming it, with words and gestures carrying significance of that power and privacy and obscurity. It does not seem like a story unfolding, but more like a history happening, and we are living through it and past it; later, we may discover what it means, if and when we discover what a life means. The knowledge of Monroe's death seems to give the movie a tragic trace. A radical contingency haunts every story of tragedy. A tragedy is about a particular death, a death neither natural nor accidental. The death is inflicted, and because it is inflicted it *need not have happened.*

Both Arthur Miller and Norman Mailer wrote 'Marilyn' into their work: they accepted the life wrapped in meaning, but they did not allow the meaning to emerge, and so they wrapped it in mystery. Possibly they sought to be respectful to Monroe, but actually they merely succeeded in being overly respectful to themselves. As late as 1969, in a relatively cool hour, Miller managed both to objectify his late wife *and* show his self-concern: he said Monroe 'was like a smashed vase. It is a beautiful thing when it is intact, but the broken pieces can cut you.' Yet what about the vase, Mr Miller? Mailer came to the conclusion that he would probably have fared no better than Miller did as Monroe's lover. The reaction typifies the self-centredness of masculine thinkers: the sadness is not for Monroe's probable suffering, but for the man's possible pain. Such writers fail to discover that we do not understand a person until we are intimate with them; but we also discover that intimacy, especially sexual intimacy, like academic knowledge, can encapsulate; less a doorway to genuine understanding than a realm in its own right, replete with rituals, habits and obsessions. We learn respect, too, for the distinctions between tacit, empathetic understanding and knowledge of fact. These drive each other out, and they do so no less in our dealings with the other's body than they do in our dealings with the other's mind. It is not simply that our needs are ambivalent; the knowledge of people to which these needs lead us is itself fraught with ambiguities and paradoxes. Whether as misfits or as Mailer, before her death or after the fall, Monroe has rarely been seen with accuracy or with respect.

There is a Russian folk story in which a man becomes perplexed by his love for a woman: he can feel the love, but he cannot find its precise cause in his lover, that aspect of her that makes him so helplessly in awe. This mystery moves him to madness: he kills his lover, and dismembers her body, seeking the object of his love. When he cannot find it, he realizes it was the very integrity and freedom of his lover that made her so special. Aware of his crime, the man takes his own life. The literary tributes to

'Marilyn' belie their motivation: this 'Marilyn' is a man's 'Marilyn', a manufactured dream of physical purity which not even Monroe was allowed to ignore. Monroe became a melancholic figure, marked and misconstrued by the masks men insisted she wear. When she resisted, she was read as 'Maggie'; when she submitted, she was read as 'Roslyn'. Never was she allowed just to be, to find a self *for* herself. Friends have said that she was acutely aware of her increasing isolation: lovers left and babies were never born. She was desperate for someone to speak with, to listen, raise an eyebrow, smile or laugh, argue with, to create a humane environment. Alone, the void opened up before her, endless and terrifying. In a typically 'masculine' way, the regrets came after the damage was done. Norman Rosten (1980: 112–13) wrote some final lines in her memory: 'she has escaped the facts and flown into myth, caught in a twilight of blended history and remembrance. She haunts us with questions that can never be answered. All beauty is mystery. What comes back to us is the smile, the desperate heart, the image that flares up and will not go away.'

Simultaneously venerated and trivialized, the image outlives the person. 'Death', said Monroe, 'is the end of everything'.
Photograph reproduced by permission of the BBC Hulton Picture Library.

7

Mourning for Marilyn

People do not die for us immediately, but remain bathed in a sort of aura of life which bears no relation to true immortality but through which they continue to occupy our thoughts in the same way as when they were alive. It is as though they were travelling abroad.

Marcel Proust

It might be kind of a relief to be finished. It's sort of like I don't know what kind of yard dash you're running, but then you're at the finish line and you sort of sigh – you've made it! But you never have – you have to start all over again.

Marilyn Monroe

'Who killed Marilyn Monroe?' said Sean O'Casey, 'that's a question.' The intrigue surrounding the death of Monroe in 1962 has been of enduring interest to her biographers, friends and followers, for over twenty-five years. Her most devoted of ex-husbands, Joe DiMaggio, chose 1982 to cancel the standing order he had started exactly two decades before for roses to be placed in front of Monroe's tomb in the cemetery. Although he refused to give a reason for his decision, friends suggested he wished Monroe to finally rest in privacy and peace. In fact, her memory has been subjected to even greater attention since that time, with the release of FBI files and the publication of new allegations concerning Monroe's relationship with John and Robert Kennedy. Gloria Steinem's *Marilyn* (1986) purports to present 'the woman who will not die', and concludes by saying, 'If we learn from the life of Marilyn Monroe, she will live on in us' (p. 180). How can Monroe continue to 'live' for us? Why do we want to say that she does live on? When asked what her own obituary would be, Monroe said: 'I'll settle for this: "Here lies Marilyn Monroe – 38-23-36'", and then she laughed. Yet her serious fear was that she would be remembered as a 'dumb blonde'. Her biographers have rarely respected this fear.

Monroe flew to Washington on 19 May 1962, in order to sing a 'Happy Birthday' tribute to President John F. Kennedy. It was a politically oriented occasion; an enviable array of entertainers had been brought together in order to attract twenty thousand prosperous Democrats willing to pay hundreds of dollars per ticket for party funds. Monroe joined such celebrities as Ella Fitzgerald, Peggy Lee, and Maria Callas. It was Monroe, nervous, shaking, sewn into her sequinned dress, who stole the show. The compère, Peter Lawford, had made Monroe's reputation for unpunctuality into a running joke, eventually announcing her as 'the *late* Marilyn Monroe'. After her performance, Kennedy said from the stage: 'Thank you. I can now retire from politics after having had, ah, "Happy Birthday" sung to me in such a sweet, wholesome way.' The following day she flew back to Los Angeles and the next Monday she was on set to film the 'nude swimming sequence' immortalized by the stills photographers present. On her thirty-sixth birthday on 1 June, she appeared with Joe DiMaggio at a baseball game, tossed the first ball and looked radiant. One week later, she was fired by her studio for her alleged unreliability. On 4 August, she rang her friend Norman Rosten in New York, arranging a date with him and his wife in September when she came east. 'We'll have a great time,' she said. 'We have to start living again, right?' Hours later, Marilyn Monroe was found dead.

It is highly probable that the first biography of Monroe one picks up will include a discussion of the 'ominous', 'prophetic' title of her unfinished movie, *Something's Got to Give,* suggesting an ironic twist of fate. There may follow a mention of the Peter Lawford '*late* Marilyn' remark. An extraordinary laziness lies behind such irresponsible inter-pretations: it is simply wrong to suggest that Monroe's early death was 'inevitable', 'fated' or even very likely. *Something's Got to Give* went into full production on 23 April 1962. Monroe's first words to the screen-writer, her old colleague Nunnally Johnson, were: 'Have you been trapped into this too?' He soon departed, but Monroe was caught in her studio contract, increasingly envious of the reports concerning Elizabeth Taylor's great autonomy on the set of *Cleopatra*. Six successive scriptwriters were hired by Fox, and Monroe would arrive each morning to find her lines rewritten and her character slightly revised. A virus infection seriously affected her ability to work; her disenchantment with the movie made her frustrated and irritable. Director George Cukor made his 'loathing' for Monroe eminently clear. After several breakdowns, Monroe was fired by Twentieth Century-Fox. Her affair with Robert

Kennedy was unstable, her marriage to Arthur Miller had been dissolved, and the press treated her with ill-concealed contempt. Asked by an interviewer whether many friends had rallied around her at this time, 'There was silence, and then, sitting very straight, eyes wide and hurt, she had answered with a tiny "No"' (in Meryman 1962). Yet Monroe was far from the passive, helpless victim; she actually began to draw upon a formidable inner strength in order to resume and improve her career.

When advisers pressured her during 1960 to have a will drawn up (a common practice for such celebrities), Monroe was offended: 'They really want to finish me off. . . Well, if it will shut everyone up, they can have their will. Won't do anyone any good, cause I'm gonna stick around for a while. . .' Monroe moved into her first 'real' home, in Brentwood, and immediately began planning its renovation. Summers (1985) makes much of the 'symbolic' motto outside the door, *Cursum Perficio* ('I am completing my journey'); Monroe was as untroubled by the 'prophetic' quality as the past and future owners. In the summer of 1962 she purchased an entire new wardrobe from Saks Fifth Avenue and Jacks of Beverly Hills. She was determined to keep the affection of her fans: 'The least I can do is give them the best they can get from me. What's the good of drawing in the next breath if all you do is let it out and draw in another?' (in Meryman 1962). Monroe made plans to present her personality in a 'more authentic' light: 'You know,' she said, 'most people really don't know me.' She held her studio especially responsible for this problem: 'I did what they said, and all it got me was a lot of abuse. Everyone's just laughing at me. I hate it. Big breasts, big ass, big deal. Can't I be anything else?' (in Luijters 1986: 10). Jean-Paul Sartre certainly wanted her to be something else – in the *Freud* movie he was writing for John Huston. There were several other film deals being discussed, and also a television role in Somerset Maugham's *Rain*. The studio was happy to use Monroe as part of a 'romance' rumour for Rock Hudson in order to help disguise his homosexuality. As a celebrity, Monroe was still highly prominent. Monroe was far from despondent: 'There's a future and I can't wait to get to it.' Pursuing her hope for acquiring more control over her image, Monroe began to tell her life-story to journalist George Barris, in order to 'set the record straight'. *Vogue* photographer Bert Stern spent several sessions taking pictures of Monroe for a prestigious feature spread: Monroe appeared as an extremely graceful figure, the sense of calm, confidence and courage was particularly impressive to the viewer.

Two days before her death, Monroe accepted a request from *Life* magazine for a major interview. Richard Meryman, who conducted the discussion, found Monroe to be a person he admired, likeable and 'very smart'. He recalls he 'felt very keenly that Marilyn knew what she was doing every minute I was there'. She insisted on few alterations to the text, apart from passages she felt might hurt her stepchildren. Monroe received considerable praise for her *Life* performance. Her distinctive personality, her humour, and her striking ability to use language with imagination and inventiveness, are evident in every line. As Meryman left her house for the last time, Monroe stood at the door and called out, 'Hey, thanks!' 'It was very touching', Meryman said, 'that little girl thing was very strong' (in Summers 1986: 374).

'Please don't make me a joke' – these were the words Monroe spoke for publication before she died. As they mourned her passing, her obituarists avoided dwelling on the rumours of foul play. A movie star's death is particularly memorable in a community dedicated to spectacle; as Billy Wilder's *Fedora* puts it, 'Endings are very important. That's what people remember. The last exit. The final close-up.' Monroe's death was reported as a straightforward 'suicide' from a drugs overdose. A person who had a great fear of choking, who found it difficult to take even one aspirin without the aid of liquid, was reported as having swallowed four dozen Nembutals and several chloral hydrate tablets. Police had found Monroe's bathroom plumbing switched off, and there were no water containers in her bedroom. It was later claimed in a news conference held by Coroner Theodore Curphey, reporting on the Suicide Investigation Team's findings, that Monroe 'took one gulp within, let's say, a period of seconds'. Walter Winchell did make a modest gesture of protest at this curious performance, mentioning a high-ranking politician 'from the east' who, he said, must be 'plagued with guilt', but no more was suggested.

Several reporters in the area, tuning into the police network on their car radios, were alerted to the news of Monroe's death. Some managed to arrive in time to observe the body being moved from the house. Photographers swarmed over the grounds, snapping through windows and half-opened doors. According to the deputy coroner's aide at the Los Angeles County Morgue, many people from all walks of life were admitted into the storage vaults where Monroe's body was kept in Crypt Number 33, and viewed the body out of curiosity (see Speriglio 1982). Two photographers successfully made their way into the morgue. One, Bud Gray, snapped a picture of Monroe's shrouded corpse whilst his colleague created a diversion. Leigh Weiner bribed an attendant and managed to take several pictures of the body, covered and uncovered. He was sufficiently callous to sell the photographs for publication, one of the images showing a toe obtruding from the crypt, an identification tag attached. The autopsy surgeon, Thomas Noguchi, and John Miner, the observer from the District Attorney's office, arrived to find the body of

Marilyn Monroe on the slab: Miner recalls, 'Tom and I had looked at thousands of bodies, but we were both very touched. We had a sense of real sadness, and the feeling that this young, young woman could stand up and get off the table at any minute' (in Summers 1986: 426).

News of Monroe's death now seeped into every possible media space, and 'personal themes' were sought. Reporters contacted Monroe's former husbands for their immediate reactions. James Dougherty simply said, 'I'm sorry.' Arthur Miller was speechless for some time, eventually saying that 'It had to happen,' and that he would avoid the funeral because Monroe was 'not really there any more'. Joe DiMaggio, who had remained close to Monroe, arrived in Los Angeles in order to make the funeral arrangements. He flatly refused to speak to the press. A friend, Harry Hall, recalls seeing DiMaggio sitting in his hotel room, head in hands, surrounded by unopened telegrams, weeping inconsolably. When he spoke at all, he shouted about Frank Sinatra and his 'Rat Pack', and made it clear that he held Robert Kennedy responsible for Monroe's death. 'It wasn't Hollywood that destroyed her,' he said, 'she was a victim of her friends.' DiMaggio and Monroe's former business manager Inez Melson, issued a prepared statement, informing the press that there would only be a small funeral, 'so that she can go to her final resting place in the quiet she has always sought.' Twenty-four mourners were invited, but other celebrities arrived with their security guards and tried to gain attendance. The waiting press photographers ran over graves as they chased after the coffin. In the mortuary, DiMaggio instructed the funeral directors, 'Be sure that none of those damned Kennedys come to the funeral.' A non-denominational pastor made a brief address, drawing on a verse from the Book of Psalms: 'How fearfully and wonderfully she was made by the Creator!' DiMaggio stood by the coffin, said, 'I love you' over and over, then bent for a final kiss. The crowds watched as Monroe was borne to the Mausoleum of Memories, and the coffin was placed into a waist-high vault. One wreath from an anonymous sender bore the text of Elizabeth Barrett Browning's sonnet that includes the line: 'I shall but love thee better after death.' For those amongst Monroe's friends who knew how deeply in need of love she had been, the words must have seemed bitterly ironic and far too late. Monroe now lies behind a marble plaque that reads: 'Marilyn Monroe 1926-1962'. In his special tribute, Lee Strasberg lamented the loss of someone whose 'career was just beginning': 'I hope that her death will stir sympathy and understanding for a sensitive artist and woman who brought joy and pleasure to the world.'

'The curtain falls. . .' began the narrator of the newsreel featuring Monroe's funeral. Tributes and evaluations from Monroe's colleagues began to appear in print. Joshua Logan described Monroe as 'one of the

most unappreciated people in the world.' The President of Twentieth
Century-Fox, Darryl Zanuck, commented: 'Nobody discovered her, she
earned her own way to stardom.' Peter Sellers, an actor who identified
with Monroe's anxiety over her off-screen identity, was described by
Jerry Juroe as 'mortified' upon hearing of her death: 'It was as if it were
the death of the closest person to him in the whole world. . . [It was] a
moment of real grief' (in Evans 1981: 192). A former co-star of Monroe's,
Jane Russell (1986) writes of an evening in 1962 when she wanted to
contact her old friend and invite her to a party. The next day, Russell and
her husband heard on the radio that Monroe had died: 'We were stunned.
If only, if only. . .' Monroe's psychiatrist friend, Ralph Greenson, was
said to be grief-stricken; he would later reflect, 'She was a poor creature,
whom I tried to help and ended up hurting.' Robert Kennedy was
socializing with friends when Monroe's death was discussed. In a 1985
BBC TV documentary, his supporter John Bates recalled that Kennedy
took the news 'rather lightly. . . It was discussed in sort of an amusing
way.'

Monroe had said to Weatherby (1976: 127): 'I've come to love that line,
"until death do us part". It always seems to go well for a time, and then
something happens. Maybe it's me.' The day after her death, her
interviewer Richard Meryman was sitting at the counter in a New York
coffee shop, idly listening to a workman talking: 'I was working on the
street outside her apartment house on 57th Street, and whenever she'd
come out she'd always say "Hi", right to me.' Then he turned to
Meryman and the other customers, declaring: 'Marilyn Monroe was the
nicest girl in the whole world.'

Monroe's fans and admirers were devastated, and many wept for the
person they had never met. Monroe's movies had grossed over £70
millions; she left £178,000 in her will, with special donations to children's
homes throughout the world (*Sunday Times* 30 December 1962). There
were reports of young female admirers taking their own lives out of
despair after Monroe's death (see *Los Angeles Times* 6–26 August 1986).
After the personal tributes there came the critical commentaries, often
written with some haste; Alistair Cooke (correspondence of 6 June 1986)
told me,

> I well remember having driven to our village on Long Island for the
> paper and returned to call the foreign editor (of *The Guardian*) and
> tell him there was no news worth writing about. He told *me* of her
> death, and I asked for an hour in which to write an obituary piece. I
> had no books to hand and wrote it out as fast as I could.

Hollywood's own resident critics were particularly uncomfortable with the story; Hedda Hopper, in an unusually honest piece of writing, confessed: 'I suppose all the sob sisters in the world will now start to go to work. In a way we're all guilty. We built her up to the skies, we loved her, but we left her lonely and afraid when she needed us most' (in Taylor 1984). Hollywood and its magazines were amongst the most disingenuous in their reactions. *Silver Screen* (December 1962) included the comment that 'Marilyn finally found tranquility in death.' A less banal appreciation, in the same issue, by Favius Friedman, argued:

> If she lived in anything, it was in the desperate knowledge that she wanted love more than anything else in the world. . . She fought until the last to forget that most of her life she was a lost child, afraid and insecure; and when peace still eluded her, at 36, she vanished with a sigh, unaware how the world would forever miss her.

The *Los Angeles Times* critic, Philip Sheuer, noted Monroe's worth whilst showing a hard-headed approach to the ways of the movie world; he concluded by saying: 'Success in show business exacts a higher price than other trades do, and Marilyn was generous in paying off. Only a Higher Power can distinguish in such matters of right and wrong, but we can wish her peace, and do.'

There came a second wave of tributes, this time from writers normally unmoved by 'popular stars'. In England, Malcolm Muggeridge devoted considerable space to his reflections on Monroe:

> Symbol or no, her real virtue was an unpretentiousness that even the most artful publicity flacks could not efface: she hated being regarded as a "thing" and was candid about her opinions, which the priggish would regard as less than respectable. Yet the nature of her death, ironically, will turn her more than ever into a "thing", a cult figure in the necrophiliac rites that already surround another symbol – James Dean. (In Taylor 1984)

Thus, even Monroe's sympathetic mourners were taking some of the unbearable poignancy out of her end by placing it within a tradition: the idols who died young – Valentino, Harlow, Dean, Monroe, company for her memory, a sense of continuity for her survivors. An example of this self-oriented thinking was Margot Fonteyn's 'tribute' to Monroe (1976: 184): 'Her beauty was so fresh and her personality so compounded of the childlike and the vulnerable that, although her early death was a tragedy, I can't help thinking it fitting that she never had to contend with the erosions that time brings to the rest of us.' Fonteyn's conception of 'tragedy' seems to be remarkably pragmatic, allowing her to defuse death's emotional charge by calling it 'fitting'. Such insensitivity fell into offensiveness with Joe Mankiewicz's suggestion that we should remember Monroe as a 'zany, wonderful dizzy blonde', for, as an ageing Monroe would have been a 'pitiful, dreadful mess', 'her death was the best thing that could have happened to her.' It is surely rather sad that the attention accorded to 'celebrities' means that such stupidities can find the authority of print.

Despite the rapid 'explanation' of Monroe's tragedy, her unique qualities continued to remind people of what they had lost. The American critic, Diana Trilling, confessed that there is 'always this shield of irony some of us raise between ourselves and any object of popular adulation, and I had made my dull point of snubbing [Monroe's] pictures.' Suddenly, Trilling was surprised by a television trailer for a Monroe programme: 'an illumination, a glow of something beyond the ordinarily human' had gone on in the room. Trilling became something of an admirer of Monroe, of her ability to 'suffer one's experience without being able to learn self-protection.' Most perceptively, Trilling (1964) argued that 'even when [Monroe] had spoken of "wanting to die" she really meant she wanted to end her suffering, not her life.' The tragedy of the death was irresolvable, for what we had lost was a person who was *not* typical. 'She was alive in a way not granted the rest of us. She communicated such a charge of vitality as altered our imagination of life, which is the job and wonder of art.'

Monroe's old employers were not slow to exploit her posthumous prominence. *The Misfits* now attracted audiences who could not avoid ascribing to the movie a deeply personal significance. Knowing that Monroe has died, and that *The Misfits* was her final screen appearance, one cannot stop oneself from seeking the signs of imminent catastrophe, in full recognition that this particular pursuit of signs only serves to heighten one's affective attachment to the figures one comes to see. The *Misfits* ends, not just the movie but also our sighting of Monroe, and the fade-out is both unusually sudden and disturbingly final. The blackness swallows up current hopes, and one feels the screen has darkened with unexpected finality, as though in fury at its lost power to enclose its content. Gable, Monroe, and later Clift, never to be seen again, only reviewed as projections of a past which will never be truly present. This poignancy bleeds through *The Misfits*, and Fox re-publicized it accordingly as 'Marilyn's last movie'. By 1963, the fragments of Monroe's *Something's Got to Give* sessions were collected in a retrospective 'tribute' movie, *Marilyn*: without sound, the moving images seemed especially haunting. Monroe, dead, began to be bigger business than when she was alive. *Playboy* boosted its sales by publishing pictures of Monroe's *Something's Got to Give* 'nude bathing scene'. Such pornographic movies as *Apple knockers and the Coke Bottle* were misleadingly advertised as starring 'a young Marilyn Monroe'. Unknown actors or 'mentors' began jostling for position as biographers started researching her life.

Monroe's image came to be circulated and exchanged within the movie industry. Fox started a Marilyn Monroe look-a-like contest, which continues to this day. Women in the movies were worked into the 'MM' mould: 'BB', 'DD' – Brigitte Bardot, Diana Dors, and whole lines of blonde-haired 'sex sirens' – Jayne Mansfield, Kim Novak, Anita Ekberg, Elke Sommers, Martine Carol, Julie Christie, Farrah Fawcett-Majors, Pia Zadora, and Bo Derek. Women who made their names on the strength of Monroe impersonations included Marilyn Marshall in nightclubs, Debbie Arnold on television, Linda Kerridge in commercials, Misty Rowe, Catherine Hicks, Tracy Gold and (to a lesser extent) Theresa Russell on film. Monroe's image was a common cultural reference: *The Apartment* (1960) can have a man say, 'Now look buddy boy, I can't pass this opportunity up – she looks just like Marilyn Monroe!' Monroe had mimicked the cartoon character Betty Boop when she sang 'I Want to be Loved By You', and when the Betty Boop cartoons were restarted she was made to parody *The Seven Year Itch* 'skirt scene'. The woman who came from nowhere is now on display everywhere.

The initial series of prints produced by Andy Warhol after Monroe's death was an indication of the potential mechanical reproducibility of a multiplicity of 'Marilyns' in the market place. Contemporary society is shot through with images from the past. Everywhere one turns, one brushes up against someone's past, an image of a past: movies, television, music, prose style, clothes, design, desires, ideologies. In refashioning the past in our own image, in tailoring the past to our own preconceptions, it is recuperated but often not respected. The 'Past' is transformed into the most disposable of consumer commodities: the people who lived it may be forgotten, the lessons which it may teach us are thought trivial. 'Marilyn' is rubber-stamped and reproducible as a marketable image: blonde hair, moist eyes, licked lips, beauty spot, soft skin like milk-filled silk, an attractive commodity for a consuming passion. Andy Warhol's *Marilyn Monroe Diptych* (1962) shows how celebrity breeds clones – thousands of signs for itself, a series without limit. It wanted to be glanced at like a television screen, not scanned like a painting. The work was morally numb, haunted by death, and disposed to treat all events as spectacle; it succeeds in making Monroe seem coarse and cheap.

As the mourning passed into manufacturing, Monroe became

distanced from certain manifestations of 'Marilyn'. Her image was used on kitsch culture postcards and posters as an embodiment of 'fun': gentlemen prefer blondes to tragedies and this 'Marilyn' was exploited for her instant attractiveness and familiarity. The repetition principle was now in action. The whole meaning of such exploitative commodities is in their unredeemed lowness, a vulgarity of outlook in every direction whatever. The most modest of hints of 'higher' influences would seriously undermine their objective. The products bearing the mark of 'Marilyn' do not contain Monroe: she is never *consumed*. During her lifetime, Monroe felt the negative effects of this consuming passion, rather in the sense of someone 'suffering from consumption' – 'everyone is always tugging at you' and crystalline celebrity hearts seem set up often only to be smashed.

Images of 'Marilyn' are set and shine into the contemporary consumer culture like colour glass in a kaleidoscope: Marilyn dolls, Marilyn lipstick, powder, make-up kits, clocks, clothes, bedsheets, playing cards, jigsaw puzzles, mugs, plates, spoons, posters, prints, party picks, writing paper – an endless spiralling of reference and cross-reference, items from an icon of the market-place. The cultural tradition is the practice of ceaselessly excavating, safeguarding, associating, violating, discarding, revising and reinventing the past. The multiform images and uses of 'Marilyn' revolve around our memory of Monroe and her distinct personality. Every year, from 1 June to 5 August (marking the anniversaries of her birth and death), there is a Marilyn Monroe Memorabilia Show and Sale held at Twentieth Century Antiques & Gallery, New York City. With remarkable regularity, a rare image of Monroe is retrieved and auctioned. Frequent re-runs of her old movies on television and in the cinema attract large audiences. Monroe has been part of our lives and fantasies for nearly four decades.

The famous white dress from *The Seven Year Itch* is perhaps the most prominent image now associated with Monroe. Fashion stores are often keen to hire models dressed as Monroe to mimic the pose and promote the product. Photography magazines such as *Zoom* (34, 1986) use the image as a cover illustration. The same image has been deployed at various times by models in perfume and alcohol advertisements. Soul singer Donna Summer dressed as Monroe for the cover of her album, *Four Seasons of Love* (1976). In 1986, it was even reported that one of the original white dresses had been stolen by a male transvestite. Gene Wilder's movie, *The Woman in Red* (1984) repeated the original pose but changed the colour to scarlet (see the discussion in chapter 1 on 'Red' and 'White' women). An episode of the television crime show *Matt Houston* featured a corrupt boss who has a fetish for Marilyn Monroe, making his lover wear the famous dress and imitate Monroe's voice and mannerisms. The whiteness of the image wraps the memory of the woman in thoughts of purity, protectiveness and playfulness. The 'white Marilyn' is the 'fun'aspect of the image, the aspect associated with sexual promise and physical perfection; such an emphasis is far removed from the artist who died in tragic circumstances, and one is thus encouraged to view 'Marilyn' with no hint of remorse: 'The glamour that Marilyn exuded is still written and talked about after her death nearly three decades later... Marilyn will remain a fascinating enigma and the personification of glamour itself... What did Marilyn Monroe have in common with Jerry Hall? Or Joan Collins with Madonna? The answer is an aura of GLAMOUR.' (*Look Now* January 1987)

'Like Marilyn Monroe': the phrase is deeply woven into the modern world, readily applicable to figures of glamour, femininity, fashion and tragic, early death. John Kobal has commented that Monroe 'stood for life. In her smile, hope was always present... she had become a legend in her own time, and in her death took her place among the myths of our century.' The legend of Monroe leads one into the bourgeois truisms of Western culture: that fame does not bring happiness; that sexuality is destructive; that Hollywood destroys its own children. In dwelling too much, too superficially, in the retrospective meaning given to Monroe's life by her final troubled hours, we deny her adulthood and agency. This emphasis, widespread and wearisome, is not merely irresponsible but also immoral. As Monroe said in her final interview: 'It's nice to be included in people's fantasies but you also like to be accepted for your own sake.' As Adorno writes of popular images (1978: 166): 'Precisely where they become controllable and objectified, where the subject believes himself entirely sure of them, memories fade like delicate wallpapers in bright sunlight.' Billy Wilder, by the mid-sixties, became disturbed at the growth of the Monroe industry: 'I am appalled by this Marilyn Monroe cult.' 'Marilyn' became a category whereby not only Monroe is positioned but also those who are identified with her: to be 'like Marilyn' is to take on a repertoire of characteristics, to be 'read' in a rapid and regularized way.

Monroe herself was often rewritten by novelists, trimming her of those qualities which resisted any stereotype. *The Symbol* (1966), written by Alvah Bessie, features a character clearly based on Monroe. After drawing quite blatantly on Monroe's biography, Bessie has his movie star reach for a full bottle of barbiturates, saying: (pp. 302–3) 'I'm gonna go to sleep now and when I wake up I wanna be rich and famous and a movie star and all the men in the world will fall in love with me and take me away from all that.' Bessie's apparent desire to 'confirm' that women are irrational beings obsessed with their ability to please men such as Bessie is crude but sadly common in the fictional treatments of Monroe. Norman

Mailer even resorted to a nursery talk, a schoolboy salacity by writing of a 'fucky Marilyn'. With the 1986 production of his 'memory play', *Strawhead,* Mailer cast his own 23-year-old daughter as Monroe. After the Strindbergian agonizing of his 1960s memoirs, Arthur Miller has waited until the late 1980s to record his modest reflections. The movie industry has been keen to appropriate 'Marilyn' for its own purposes: *Tootsie* (1982) is merely one of many movie references to Monroe, featuring both her image and an associated theme (from *Some Like It Hot*). Monroe's movie biographies have been significant for their silence on the role of Hollywood in her problems: thus, in *The Goddess* (1958), *Goodbye Norma Jean* (1976) and *Marilyn: The Untold Story* (1980), one finds the predictable thesis that Monroe was doomed long before Hollywood began 'irritating' her.

Two British directors have made movies that depict Monroe as the object of mass adulation. Ken Russell's work, beneath a specious but plausible artistic pretext, constitutes a kind of cultural pornography, both degraded and degrading. Russell's *Tommy* (1977) contains a scene in which the deaf, dumb and blind boy is taken by his mother to a cathedral covered in pop cultural icons, its walls furnished with huge film stills featuring Marilyn Monroe. As the lyric 'See me, feel me, touch me, heal me' is sung, a huge effigy of Monroe, dressed in *The Seven Year Itch* costume, is carried into church by her many acolytes, all masked as Monroe. The scene is a particularly cynical vision of Monroe-as-threat, of sexuality as unsettling and of fandom as fatal. People touch their effigy to be healed. In a struggle, Tommy breaks away and the statue smashes into pieces, showing the emptiness of its promise. As with Arthur Miller's 'beautiful vase' reference to Monroe, Russell sees her as an object for his possible pleasure or pain; he reveals a disturbingly unbalanced perception of Monroe's significance. Terry Johnson's play, *Insignificance,* was adapted by him for Nicolas Roeg's movie of 1985. The story is set in 1953, during the making of *The Seven Year Itch,* where four characters (based on Monroe, Einstein, DiMaggio and Senator McCarthy) are brought together in a hotel room to reveal their fears and obsessions. An aggressive DiMaggio shouts: 'If I want to see my wife, I go to the movies. If I want to see her in her underwear, I just go down the street with all the other guys.' 'Marilyn', gazing out at her image illuminated by the neon in the night sky, sighs: 'I wish they'd switch me off.' She looks to Einstein, saying: 'I just thought that we could come together in the middle of all this – just for an hour or so.' Only when the clock stops does time come alive. Each mythic figure in this eerie milieu is, and remains, enigmatic; the ever-changing density of the story, rolling along gathering momentum, seems to go nowhere yet arrives everywhere as it explodes in a shower of fractured illumination. Roeg, who courts ambiguity in his work, cannot escape the charge that he depicts Monroe in a condescending light: the implicit 'joke' is that Monroe gets on well with Albert Einstein, the interest is in celebrities rather than people.

During the 1960s, when celebrity culture shifted from the stars of Hollywood to the Beatles and pop music, Monroe was studied as a recent example of the mass idol. John Lennon recalled: '[The Beatles] didn't have any hope just because we were famous. . . You see, Marilyn Monroe and all the other people, they had everything the Beatles had, but it's no answer' (in Yorke 1970). Lennon selected Monroe and other famous figures when the Beatles collected sixty-two celebrity images for Peter Blake's surreal collage cover for *Sergeant Pepper's Lonely Hearts Club Band* (1967). When the rock stars met with early deaths, including Brian Jones, Janis Joplin and Jimi Hendrix, they were immediately related to the example of Monroe. A survivor, Bob Dylan, told the interviewer Cameron Crowe: 'People like to talk about the new image of America, but to me it's still the old one – Marlon Brando, James Dean, Marilyn Monroe. . . I like to stay a part of that stuff that don't change' (in Shelton 1986: 497). Once asked whom he would like to interview, Dylan replied: 'A lot of people who aren't alive: Hank Williams, Apollinaire, Joseph from the Bible, Marilyn Monroe, John F. Kennedy, Mohammed, Paul the Apostle, maybe John Wilkes Booth, maybe Gogol. I'd like to interview people who died leaving a great unsolved mess behind, who left people for ages to do nothing but speculate.'

No single celebrity has been the object of such intense interest from popular songwriters as Marilyn Monroe. Elton John and Bernie Taupin have referred to Monroe in the songs 'Candle in the Wind' (1974) and 'Wrap Her Up' (1985). Leon Russell's 'Elvis and Marilyn' (1978) offered the audience a mawkish mass cultural necropolis. Other songs referring to Monroe include David Bowie's 'Jean Genie' (1973), B.A.D.'s 'E = MC²' (1985), and Everything But the Girl's 'Sugar Finney' (1986). Tom Waits, in his sardonic stories of American anomie, sings of Monroe in 'A Sweet Little Bullet' (from the album *Blue Valentine*, 1978):

> I hear the sirens in the street
> All my dreams are made of chrome
> I have no way to get back home
> I'd rather die before I wake
> Like Marilyn Monroe

Waits, during 'Jitterbug Boy', sings: 'Buddy I've done it all/Because I've slept with the lions/And Marilyn Monroe' (from the album *Small*

Change, 1976). Boy George, discussing his period of drug addiction during 1986, cited Monroe as an earlier victim of showbusiness pressures. Peter Robinson changed his name to 'Marilyn' and dressed as Monroe before becoming a singer: he told reporters, 'I get really upset when I think about her sometimes.'

Time Out (11–18 February 1987) featured an interview with Peter Stack, who appears in revue as 'Dead Marilyn'. In his performance as the dead star, she appears clawing her way out from a grave, her white dress torn and her pasty flesh shredded. Stack explains his necromantic exhibitionism:

> I always used to do Marilyn from the grave at Hallowe'en parties. After this became an annual event the clubs started throwing "Dead Marilyn Hallowe'en Parties" and expecting me to show up. . . Every night [Monroe] would come to me in my sleep and give me vital information about herself and inspire me to feel free to tell her story.

When 'Dead Marilyn' performed during 1986 at Club DV8 in San Francisco, there were fifty Monroe look-alikes in the audience, many of whom wore mud-flecked blonde wigs, torn white dresses and carried lit candles. 'Dead Marilyn' arrived in a hearse and was carried to the stage in a coffin. 'Outrageous' and 'controversial', 'Dead Marilyn' was banned from appearing at certain venues, but attracted a relatively loyal and enthusiastic following on both sides of the Atlantic. 'Now nobody,' he boasts, 'does it deader.' He fed off the need for further exposure of Monroe's 'secret lives', the fascination with her final hours alive, the hours following her death. The performance, intended by Stack as a 'tribute', exploits the memory of Monroe in the most tasteless way imaginable, eschewing compassion and careful argument in favour of the shock-value of the decomposing figure, serving anyone's ends except Monroe's. It represents a shallow but persistent urge to avoid the fact of a death, by playing with images, placing images in place of the person, in the place where we are and she was, by giving us 'another way of looking at Marilyn', another way of avoiding her absence. The persistence of the image, no matter how distorted, preserves the semblance of normality: switch it on, turn it up, print it, set it, into sight, out of mind, anything to fill the void and empty the anxiety.

Two female pop stars of the late 1970s and 1980s, Deborah Harry and Madonna, made explicit reference to their resemblance to Monroe. Harry dyed her hair blonde and formed the group Blondie. She allowed it to be quoted that she regarded Monroe as her 'real mother'. Madonna used Monroe's myth but without the aura of unavailability attached to the original image. Madonna said to reporters: 'I can feel the spirit of Marilyn Monroe within me. We are one' (*Sunday Mirror* 14 December 1986). She made promotional videos based on dance routines from *Gentlemen Prefer Blondes*. Both Harry and Madonna soon started to make movies. When their fame reached a peak, pornographic pictures they had posed for when struggling to start their careers were exhumed and exhibited. Throughout their careers, the image of Marilyn Monroe encircled their celebrity. Amidst the frenzy of fame, one loses sense of historical proportion, unable realistically to assess either its past or future. The modern star performer inhabits a celebrity world with fellow famous figures with the same histories of success, the same cultural references. Referring to Monroe, these figures refer themselves to a tradition and a narrative of stardom.

Monroe's memory was constantly rewritten by writers of articles, memoirs and biographies. The majority follow the prescription by Miss Prism: 'The good ended happily, and the bad unhappily. That is what Fiction means.' Old lovers and colleagues sought to position Monroe as an item in their own biographies: her first husband, James Dougherty, wrote of a young Norma Jeane who found him strong, wise, the centre of her universe. Arthur Miller, agonizing over his feelings of guilt, rewrote Monroe as a vicious ego-maniac. His long-awaited memoires proved fairly unilluminating as regards his relationship with Monroe. Monroe's former maid, Lena Pepitone, collaborated with William Stadiem to write a dubious account with dirty fingerprints, soiled sheets, and shallow sensationalism. Biographers from Zolotow to Summers have charted the arc of her life and accorded it the aura of inevitability. Theweleit (1987: 370) has argued:

> America's male writers pounced on the dead body of Marilyn Monroe
> so they could make a new image out of the image that had been
> rendered unmistakably human by her suicide. They tried to add the
> image of a "soulful" Marilyn to the image of a physical one. That was
> a necessary step, in that she had just notified them that she was
> cancelling the contract for patriarchal dominance.

Writing in *A Grief Observed* (1985) about the death of his wife, C. S. Lewis comes to realize how much of his fear and sadness is for *him* rather than for her absence, and he rewrites her (p. 18):

> Slowly, quietly, like snow-flakes – like the small flakes that come when
> it is going to snow all night – little flakes of me, my impressions, my
> selections, are settling down on the image of her. The real shape will
> be quite hidden in the end. . . . The rough, sharp, cleansing tang of her
> otherness is gone.

Men mourning Monroe worked to close the gap between themselves and her own identity, to take her, to remake her, and send her back out into the world as the mythic 'Marilyn' with the male signature. The death of the star allowed the symbol to assume a harder outline: bold, bright, lip-glossed and loveable, a Mailer-made woman for men's amusement. Unlike Grace Kelly or Marlene Dietrich or Greta Garbo, Marilyn Monroe was available for public adulation: she made a special effort to seem 'one of the people'. In death, her very public image was eminently accessible, causing the kind of disrespect men feel for women they can 'possess'.

The 1960s was a decade in which the available images and anecdotes concerning Monroe were exploited in every possible way. When the resources seemed near to exhaustion, the memory was mined for further possibilities, producing an increasingly dubious excavation of Monroe's life history. Kenneth Anger's *Hollywood Babylon* series, featuring ghoulish revelations about dead celebrities, devoted considerable space to Monroe. The general trend was away from Monroe's life to the circumstances of her death. The material has always been compelling: the facts of her final days leap like a knife from the memory. The increased popularity of 'investigative journalism' during the 1970s, especially following Watergate, encouraged an often merciless examination of the Monroe myth. Shadows of meaning, once cast by flickering fires, were suddenly pinned down by the beam of electric light bulbs: what was open-ended became crystal clear and closed. Investigative journalism is rather like an old-fashioned whodunit: it assumes that there is a body in the library. There always has to be a hero (usually the investigative journalist) and a source of evil. Most of what it reports as 'fact' is rumour and conjecture, especially in the story's early stages of development, and it relies for its effectiveness upon the accretion of detail that will encourage what the novelist must encourage: the willing suspension of disbelief. Investigative journalists have written extensively on the 'probable' scene in Monroe's bedroom as the life began to leave her. One of the most extraordinary conceits of these professional voyeurs is their apparent conviction that, by discovering the furniture of a room they determine the furnishings of a mind. Through interviewing people with the implicit expectation that an opened mouth means a mind made up, these journalists can exhibit as little tact or taste as those photographers who trampled over graves at Monroe's funeral. As Wilde once said, 'Those who find ugly meanings in beautiful things are corrupt without being charming.'

The charmless chasing after 'the secret lives of Monroe' was initiated before her death. In 1977, a man repairing the roof of Monroe's old home had his foot push through the tiles, which had rotted away over the years.

Beneath it was a shallow crawl-space, wherein he found wiring and some old rusted relay transmitters. It was estimated that the equipment had been under the tiles for around fifteen to twenty years. Monroe's repeated insistence during her final year that she was under surveillance has been treated by many biographers as evidence of 'paranoia'. She was forced to make most of her personal calls from nearby pay telephones, carrying with her a large purse full of small change. Her increasing anxiety and nervousness was simply 'explained' as one more instance of psychological instability. In fact, J. Edgar Hoover is reported to have informed the Kennedys of the 'existence of tapes containing compromising conversations' between themselves and Monroe, and James Hoffa was said to be planning to use the tapes against Robert Kennedy before the latter's assassination (see Speriglio 1982 and Summers 1985).

In 1964, a booklet was published by Frank Capell, entitled *The Strange Death of Marilyn Monroe*, which first pricked the public interest in the Kennedys and Monroe. Capell was a right-wing activist inspired by his loyalty to the Mafia, but his allegations were sufficiently damaging to require J. Edgar Hoover to protect the Attorney General. All writers who print material on the subject are noted in the voluminous FBI file on Monroe. The New York home of the leading organizer of illegal electronic surveillance in the USA, Bernard Spindel, was raided by the police during 1966. A report in the *New York Times* (21 December 1966) noted: 'In an affidavit submitted to the court, Bernard Spindel asserted that some of the seized material contained tapes and evidence concerning the circumstances surrounding the death of Marilyn Monroe, which strongly suggests that the officially reported circumstances of her demise were erroneous.' Spindel, a highly experienced spy, had even hidden bugging devices in his own home. Thus, the police raid was captured on tape, and one conversation can be heard in which someone says, 'What do the Marilyn tapes have to do with Bobby Kennedy?'

The first book to make a serious, sustained 'conspiracy' charge, was *The Curious Death of Marilyn Monroe* (1974) by Robert Slatzer. Slatzer, who claimed to have been married to Monroe for just five days, argued that she had been the victim of a murder cover-up. His case was taken up by Milo Speriglio, director of Nick Harris Detectives in Los Angeles. The two men have since been the principal source of information for journalists such as Anthony Summers. During 1974, they petitioned the Los Angeles County Grand Jury for an investigation into the death of Monroe. The response was negative. The following year, armed with even more evidence, they were again turned down. The Senate Intelligence Committee, formed in 1975 to investigate the alleged subversive actions of the CIA, exacerbated the interest in the conspiracy thesis by disclosing details of Judith Exner's affairs with John F. Kennedy and Mafia boss Sam Giancana. Woodward and Bernstein's *The Final Days* (1976) has Richard Nixon state that he was curious as to what Marilyn Monroe had to do with national security. Speriglio decided to publish the evidence he had accumulated in *Marilyn Monroe: Murder Cover-up* (1982). In August of that year, the twentieth anniversary of Monroe's death, *US Magazine* ran a feature on Speriglio's contention that Monroe had been murdered. He held a press conference at the Greater Los Angeles Press Club, and told the largest audience in the club's history that he was demanding an official investigation into Monroe's death.

Norman Mailer, whose *Marilyn* (1973) had implicated the Kennedys without providing any substantive evidence, remained eager to see the subject pursued. He tells me (correspondence of 6 January 1987) that he contacted Anthony Summers, telling him, 'I was not happy with the research I did on Marilyn's death, and indeed, I was the one who first tried to get him interested in the question of just how she died.' Summers says that his determination 'to tackle the death issue' led to his 'fresh and stimulating book', *Goddess: The Secret Lives of Marilyn Monroe* (1985). The study, whilst providing valuable new facts which cannot be ignored by future biographers, is seriously shallow in its search for ever more sensational stories. Summers, a 'trained investigative journalist', cannot resist trying to 'find Marilyn' – even to the extent of entering the morgue. It is a particularly great misfortune that Summers should find it necessary for his narrative to include a photograph of Monroe's corpse. It is not so much what she looks like that appals, but rather the feeling that this is the ultimate intrusion, on our part as well as on the author's. It is a needless gesture from the author, a cynical sign, one which reflects very badly on his profession.

Summers takes many of his key points from Speriglio, who himself entered the 'celebrity journalist' world by collaborating with Steven Chain on *The Marilyn Conspiracy* (1986). The sleeve blurb blurts out: 'After thirteen years of relentless detective work, Milo Speriglio has tracked down the tragic, terrifying truth about the sinister death of Hollywood's most unforgettable love goddess.' In case we miss the point, the bold, blood-red type bellows: 'AT LAST, THE TRUTH WILL BE TOLD!' The author, the man who tells the truth, opens by speaking directly to you, dear reader: 'My name is Milo Speriglio.' You are suddenly in a smoke-filled world of dark, doom and death. 'Detectives are people who work with puzzles,' he obligingly explains. 'I've been working on a particularly intriguing puzzle ... the death of Marilyn Monroe.' Sentences are spat out with Bogart bluntness, the paragraphs are set down and stubbed out like cheap cigarettes worth one perfunctory puff. 'When a detective tries to unravel a crime,' snaps Speriglio, 'he begins at the beginning. He looks for a motive.' The reader is thus left in no doubt that the sleuth started with an end in mind, and found it: the book 'will show beyond the shadow of a doubt that Marilyn Monroe did not commit suicide'. Although Speriglio does not live up to his promise, he does provide sufficiently striking revelations to stimulate further research into the death of Monroe – a death which seems to cause more tension than tears. The detective neither identifies with the victim nor the criminal: rather he unravels the puzzle largely in order to enjoy the process of unravelling. In a milieu robbed of genuine human emotion, the feeling which the decisive action of the investigative account answers is simply that of tension. 'They'd all like sort of a chunk of you,' said Monroe. Represented, repossessed, reinvestigated, resented, reviled, revived, revered, reviewed: 'it's like "rrr do this, rrr do that. . ." but you do want to stay intact.' The investigative journalists have been digging up the memories of Monroe for over two decades, and rarely have they tempered their approach with respect for their subject. As Karl Kraus remarked, 'Lord, forgive them, for they *know* what they do!'

For all its black and white, grainy aura and its superficial resemblance to some of the more regrettable crime movies, this story remains a thoroughly modern tragedy. We are concerned with a concocted and corrupt 'Camelot' in which 'how to handle a woman' receives a brief but brutal reply. We are confronted with images of 'star' politicians with sex appeal and a love goddess who 'lov'd not wisely but too well'. We learn not of shady, solitary snoopers, but rather of a highly disciplined organization with professional expertise in the practice of electronic surveillance and wiretapping. Death circulates in the plot and its trace seems to leave no one untainted. It seems particularly ironic that John F. Kennedy, the first media president, should have been tracked and possibly trapped by the wirings and wiles of the self-same media technology.

The mass media feed off figures like Monroe and the Kennedys. It is not so much a symbiotic as a parasitic relationship, and a prudent parasite does not kill its hosts. So long as the controversy is contested by two opposing camps populated by Kennedy cohorts and Monroe mourners, the 'conspiracy' remains a commodity of some value to the media – regardless of individual loyalties or personal tastes. If, however, the case is considered on more objective grounds, by those who object to the deliberate inconsistencies and the arrogant evasiveness of people in positions of national and international power, then it becomes an issue of some urgency. Arthur M. Schlesinger's official biography of Robert Kennedy would have been an eminently suitable opportunity for putting the 'Marilyn connection' to rest; in fact, the subject was given a single-page mention in such a way as to arouse suspicions rather than squash them. It seems intolerable that the image of Monroe, after her death, has been repeatedly exchanged between rival groups, each with their own vested interest in the conspiracy theory. The whole genre is so depressing not merely because of the frequent lapses into vulgarity, or for the sense of voyeurism in the analyses, but also because so little of what is suggested strikes one as incredible.

Benjamin (1979: 99) observed: 'A bearer of news of death appears to himself as very important. His feeling – even against all reason – makes him a messenger from the realm of the dead.' Writers on Monroe have become, in John Berger's words (1984: 240), 'Death's secretaries': 'It is Death who hands them the file. The file is full of sheets of uniformly black paper but they have eyes for reading them and from this file they construct a story for the living.' Death, real death, is when the witness dies. As is said in Stoppard's *Rosencrantz and Guildenstern are Dead* (1967): 'Dying is not romantic, and death is not a game which will soon be over. . . Death is not anything . . . It's the absence of presence, nothing more . . . the endless time of never coming back . . . a gap you can't see, and when the wind blows through it, it makes no sound.'

We desperately cling to the memory of Monroe, as though to loosen our grasp would send her over the edge and into oblivion. We grieve as we grieve for the still-born child: we project (in our minds) a future space into which that child would fit. This grief is overwhelmed by the gory details exhumed by the investigative journalist. What are we attempting when we focus on Monroe's death, be it through biographical narration or Anthony Summers' corpse picture? We in effect 'screen off' the sight of our own mortality, for even the thought of Monroe's absence fills the space between ourselves and her loss. As grief comes to displace pleasure, one can no longer ignore one's predilection for certain photographs related to one's personal vision of catastrophe. The Greeks believed that they entered death backwards – what they confronted was their past. An image of Monroe becomes a screen which simultaneously reveals and conceals. What it reveals, what we 'find' in it, is an image of the body. What it conceals, what we experience as a loss, is the being of Monroe.

We are subjected to a succession of 'previously unpublished' pictures of Monroe, some worthwhile and some not, as though beneath each new image one may find a human being, as if buried under a layer of snow. Each successive photograph presented obscures the remembrance of death that is covered in each individual image. Someone's present is shown as eternal, snatched to be sheltered from death. Summers' use of the image of Monroe's corpse is ignorant of the irony surrounding death: the horror that one cannot say anything about the death of Monroe, apart from the fact that one has 'nothing to say'. The corpse fills the space where Monroe no longer is, it discourages us from fully appreciating her absence. The image is *violent,* for it fills the sight by force, and nothing in it can be refused or revised.

The imagination cannot accommodate death: its plenitude must resist that traumatic moment in which, confronted with lack and difference, I recognize that I can die because the world is not dependent upon me for its existence. The dominant association is with our own transience. 'MM', marks of dissolution figure as *memento mori,* compelling reminders of death's imminence: 'As I am, so you shall be.' The indifference of the Monroe industry to the real tragedy of her demise is linked to the insensitive, myopic approach to deeply felt values. As Bellow says, 'Death is the dark backing a mirror needs if we are to see anything.' The 'Marilyn lives forever' theme is fuelled by fear and anxiety, a resistance to seeing the end of Monroe for fear of seeing the end of oneself. When one *does* face up to her loss, one finds the sadness expressed so well by Beckett (1983: 257): 'no questions now of ever finding again that white speck lost in whiteness, to see if they still lie still in the stress of that storm, or of a worse storm, or in the black dark for good, or the great whiteness unchanging.'

Monroe, her white dress waving about her legs, her blonde hair falling over her ears, is still a celebrity. Critics and fans are tugging at her still: wanting flesh in a hundred pieces, thousands of words with millions of pictures. During Monroe's last movie, *The Misfits,* director John Huston saw her holding some dice in her hands and hesitating. He said, 'don't think honey, throw.' The remark is typical of Hollywood's self-conscious naivety: the community's refusal to reflect, to explain, to examine its own creativity made it the most productive of movie centres, but also one of the most insensitive and cruel. When Monroe's biographer Fred Lawrence Guiles writes that 'her early death was inevitable,' he expresses the secret, perhaps unconscious, wish of anyone struggling to cope with guilt and remorse: namely, *make me forget how much it hurts to remember.* One of Monroe's more sensitive commentators, W. J. Weatherby (1976: 214), noted the problem of his painful memories when he wrote an appreciation of Monroe for the *Guardian:* 'Luckily, the newspaper's impersonal style meant that I didn't have to deal with my own feelings.' It was a mixed blessing for him, as the most precious and pertinent thoughts rubbed against the limits of his language: 'I wished I could have given everyone a picture of her smiling through the cab window that last time I saw her.'

Weatherby was forced to come to terms with his own memory of Monroe after he experienced the constant exploitation of her image and her individuality (p. 225):

The puzzle about her death was supplanting interest in her herself and
in her life. Marilyn Monroe, who in life had often been the victim of
attempts to cheapen her into a dumb blonde, was now in danger of
being reduced to a bit player in the continuing Kennedy saga or a
murder victim in a murky Hollywood mystery. . . [S]peculation about
the answer threatened to turn her complex life story into a mere
whodunit.

After the success of *Some Like It Hot,* Monroe had worried that she
would never be considered for 'serious' roles: 'That's it! I'm stuck. I'm a
dumb blonde for ever now. I've ruined everything for myself!' Although
she disproved the comment with *The Misfits,* after her death she was cast
as the dumb blonde in a new plot. The trail of Monroe's final hours has
been lost in the faded tracks of those individuals and groups who twisted
and turned until time obscured their involvement. The investigators have
sought to reduce Monroe's life to the 'hard facts', like salt wearing down
to the bone. From the results of this process it now seems unlikely that
Monroe simply decided to take her own life, and it is thus quite tempting
to favour one of the more sinister theories. What is certain is that by
choosing one theory, one keeps in motion the speculation: 'You're only
living when you can watch something die!' Weatherby has told me of the
'investigative' books: 'They don't mean much to me. They don't succeed
in bringing Marilyn alive except as a journalistic object, and their
much-touted marshalling of facts is mainly trivial detail and, where they
try to answer the questions relating to her death, a mosaic of facts and
speculation cleverly concealed as facts by putting it in someone's mouth.'

We are offered books which purport to reveal Monroe in her entirety:
pictures that cover her every angle, expression and mannerism, stories
that slice up her life and serve it on the page for your perusal. We are
encouraged to take complete possession of Monroe, as though the process
of our eyes meeting the page acts as a kiss of life to the memory of the
person. Yet this is far removed from genuine compassion. We love only
what we do not entirely possess. What we accept, we cannot possess:
'Written kisses do not reach their destination, the fantoms drink them on
the way' (Kafka).

Monroe once said, 'I knew I belonged to the audience, that I belonged
to the world, not because of my talent, not even because I was beautiful,
but because I never belonged to any one individual.' With her death, her
memory became public property, generalized and thereby compromised.
In orienting ourselves towards the future, we upset the past. In trying to
understand a person, we sometimes rewrite the personality. Popular
biography is an adversary which goes on absorbing and negating human

initiative, making the life into 'a good read'. Surely, some people *are* fascinated by stories of Monroe's mysterious death – yet when this becomes a fixation it also becomes unfeeling. As Steiner (1984: 31) has suggested, the eventual 'readability' of lives can cause one to lose a sense of the absence behind the lines:

> The capacity for imaginative reflex, for moral risk in any human being
> is not limitless; on the contrary, it can be rapidly absorbed by fictions,
> and thus the cry in the poem may come to sound louder, more urgent,
> more real than the cry in the street outside. The death in the novel
> may move us more potently than the death in the next room. Thus
> there may be a covert, betraying link between the cultivation of
> aesthetic response and the potential of personal inhumanity.

In Agatha Christie's whodunits, the reader is offered primarily an intellectual satisfaction. The basic formula is this: a murder occurs: many are suspected; all but one suspect are eliminated; the murderer is arrested or dies. The crime in this story is merely the means to an end, which is detection. In 'The Body in the Library', the object which receives the least concern from the reader is the body itself: no longer a person, only a victim, unseen and ultimately uncared for. Love celebrates the unique, death destroys it. One of the most disappointing features of the public fascination with *'who killed Marilyn'* is the signal lack of sadness, indeed, the apparent relish for the subject.

The passage of time, notes Proust (1983, vol. 3: 1106), threatens our emotional commitment:

> For after death Time withdraws from the body, and the memories, so
> indifferent, grown so pale, are effaced in her who no longer exists, as
> they soon will be in the lover whom for a while they continue to
> torment but in whom before long they will perish, once the desire that
> owed its inspiration to a living body is no longer there to sustain them.

Monroe's absence has caused her audience an anxiety which, for some, is deeply unsettling. Biographers try to put Monroe back together with pictures, paper and paste, but the result is always unsatisfactory. Where the body is absent, a heartbeat is far away: there is no body beneath the page we are holding. A text is not a tomb, the words on the cover do not function in the same way as does an epitaph. However, if we imagine it does, the book will indeed become a kind of tomb. What happens when we closely study the page of the book containing Monroe's own comments is that we react to the tone, a tone by turns hopeful and pathetic. The tone forces us to respond to the combination of the naive and the prophetic, at the uneasy mixture of the eternal and the evanescent. Monroe's poetic fragment, containing the line, 'I feel life coming closer,' conveys not a meaning, not a message, but a person: Marilyn Monroe. Here we touch her longings, her ineluctable privacy.

One comes to care for the body in the library, to move the body from the library into the heart, when one allows oneself and makes oneself see the irreplaceable qualities that now only exist in memory. As Weatherby (1976: 228) came to reflect about Monroe:

> Whichever way she died, there were always murderers in her life, trying to kill off 999 of her selves and leave her as a dumb blonde. They do the same today with her memory. Of the selves I saw ... the most memorable for me was the famous movie star who still had a genuine human interest in a wino in the street, concern for a sparrow among the pigeons. I have known many famous people, but none like that.

Past performances by Monroe, that continue to provide us with pleasure or a knowledge of the way things are, equally provide us with a sense of the performer's position toward this revelation – a position, perhaps, of conviction, concern, compassion, of confusion or ironic amusement, of scorn or shock or rage or loss. The projection of Monroe, in movies, quotations and prints, at certain times in certain places seems convincingly responsible. In the still vivid vulnerability of that laugh, in the shake of the head, in the phrase describing a feeling, in the light that lurks in the eyes, the knowledge of mortality has space to live. The past arises, the future falls, the memory of Monroe grows more binding, more brilliant.

Who mourns for Marilyn Monroe? Those who feel some empathy for her also experience an anxiety over her absence. Schopenhauer remarked, 'The deep pain that is felt at the death of every friendly soul arises from the feeling that there is in every individual something which is inexpressible, peculiar to [her] alone, and is, therefore, absolutely and *irretrievably* lost.' It is this sense of loss that should inform the writings on Monroe, rather than the common attempt to make her death mean something 'positive' for us. The fact that something feels unbearable does not for that reason make it untrue – it simply means that we cannot bear to believe it to be so unremittingly hopeless. We must resist the obsession with 'what happend to Marilyn' on that night of 4 August 1962. 'What *really* happened?' What really happened? She died. Surely that is enough.

'It's often just enough to be with someone. I don't even need to touch them. Not even talk. A feeling passes between you both. You're not alone.'

8

The Body in the Library

*I don't think I've ever met a writer I'd like as my judge. They observe
people, but often they don't feel them... But I think you've got to love
people, all kinds of people, to be able to have an opinion about them
that's worth anything... We can try to be better, and part of trying is
not to condemn other people.*

Marilyn Monroe

*When you write the biography of a friend, you must do it as if you
were taking* revenge *for him.*

Gustave Flaubert

'Who is she – who does she think she is, Marilyn Monroe?' At a party in
1962, Marilyn Monroe signed the guest register: under 'residence' she
wrote 'nowhere' – three days later, she was dead. Much was made of that
final scene with her outstretched hand tightly clutching the telephone
receiver; in the years since her death, biographers have been trying to
resume communication with the body in the library, dialling every
number, searching every name, trying every address, yet finding no
answer, no reply at all. Hollywood not only sells its product, it also
markets its guilt; the memory of Monroe pricks the collective conscience
whilst picking the profits. As an embodiment of the contradictions of a
particular feminine stereotype, Monroe reminds us of the death-wish
hidden in narcissism, the darker side of glamour. As a symbol from a
culture wherein politics, crime and corruption are common bedfellows,
Monroe reveals the tensions between innocence and exploitation. As a
subject for a biography, Marilyn Monroe draws out the writer's most
intimate feelings and fantasies, forcing us to reflect upon the simple
conceit of judging another person.

Monroe, as Ella Fitzgerald remarked, 'was an unusual woman – a little ahead of her times. And she didn't know it.' Monroe surely *was* an unusual person – that is the reason for resisting reducing her to a type, be it a 'sex symbol', a 'social construction', a 'victim' or, indeed, a 'joke'. As well as being an unusual person, Monroe was an exceptional star: popular, complicated, creative, desirable, loveable and lasting. Stars encourage us to reflect on the separation of ourselves into public and private selves, producers and consumers, extroverts and introverts: this is a key reason why stars matter to us. Stars have images, images which have histories, histories which outlive the star's own lifetime. The stars who died famous and famously, mass idols mass-mourned, like the Kennedys and Monroe, are kept alive in memory by the untimeliness of their endings, their impact upon *our* endings. To feel for Monroe, to feel *as* Monroe, we make her 'live' again, we cancel (momentarily) the reality of her death – the person appears in our library.

Pete Martin, an early biographer of Monroe, believes that she 'was born afraid. She never got over it. In the end fear killed her.' Yet the fear which killed Monroe was not entirely, or even mainly, her own. The roots of the officially bleached denouement are located in a darker, deeper context, a place somewhere between mass culture and mob murder. When the Kennedys began their rise to power, politics and popular culture were crossing over, causing Norman Mailer to write: 'America's politics would now be also America's favourite movie.' The president and the showgirl – a dream ticket with a nightmare finish. 'Fame will go by,' Monroe said, 'and, so long, I've had you, Fame. If it goes by, I've always known it was fickle. So at least it's something I've experienced, but that's not where I live' (in Meryman 1962). Where she did live, she found insecurity and insensitivity.

Monroe was a winner in a competitive society, a star system, a capitalist system, and she was flaunted so that the system was seen to work. The system worked on her as a *name*: 'MM'. 'Marilyn' shared similarities with other stars which separated her from other people. Monroe shared similarities with other people which separated her from other stars. To have made a name for herself, to have developed an image and an identity in a nation of immigrants, assuaged the trauma of being uprooted. Perhaps America needs stars more than it needs movies. John Hinckley, explaining his attempt to shoot President Reagan, described it as 'a movie starring *me*'. The famous keep alive the romance of individualism and social mobility, as celebrity is 'democracy's' vindictive triumph over equality: the body in the gutter, the name in the stars. Each constellation contains images which blend a sense of felt reality with the invented, not to say fantastical. Each falling star is viewed as a sacrifice that brings renewed fertility in the shape of someone still richer and more glamorous: goodbye last year's model, goodbye Norma Jeane (we know you had to go) – yet there has never been another Marilyn Monroe. Adoration can lead to assassination – identification forcing the devotees to usurp the object of their devotion: Mark Chapman's 'love' for John Lennon leading to his murder of his idol. Monroe's fans were 'tugging at her', literally pulling at her skin and hair; she withdrew, became more distant from those she cared for, and she would never again find a way of bridging the gap. After her death, the gap was obscured by a never-ending stairway of stories purporting to lead the reader, in easy steps, to the object of desire (the body in the library). Scanning the biographies of Monroe, there seems a national need to prove that this woman, who had once been worshipped, was not simply no better than, but actually *far worse* than, anyone else. No one is allowed to be special for too long. 'All we demanded,' she once said of her fellow stars, 'was our right to twinkle.'

Monroe starred in a mere eleven movies during a ten-year period, but the movies were merely one part of the Monroe myth. Despite her wretched childhood, she was a successful woman; although portraying the definitive 'dumb blonde', she was clearly a witty comedienne; her sometimes blatant sexuality was undermined by a childlike playfulness which transformed eroticism into physical celebration. 'Monroe's life and career', notes Rollyson (1986: 6) 'have to be seen in terms of her constant inventiveness, of her attempts at artistry in all things.' He explains: 'As a person, an actress, and a mythic figure she had different kinds of work to do, but each kind called upon what the actress knew best: role making.' As Mrs Joe DiMaggio, Monroe was sometimes beaten for displaying her sexuality; as Mrs Arthur Miller, Monroe became an interesting case-study in neurosis; as the supposed victim of a murder plot, Monroe became a thoroughly helpless figure in someone else's prurient intrigue. Some suggested that Monroe's sole lasting love affair was with the camera. It is probably true that at least this affair was not so exploitative of her emotional life. It is certainly true that she never attracted a more ardent admirer. With her premature passing, the process of image vampirism was complete: the life had been drained from the body and stored in the screen.

Norman Rosten still possesses a postcard Monroe sent him of an American Airlines jet in flight, and the written message: 'Guess where I am? Love, Marilyn' (Rosten 1980: 113). Monroe is now everywhere yet nowhere: her image is on walls, in movies, in books – all after-images, images obscuring the fact of her permanent absence. For the movie industry, to place the consumer of its products in a constant position of desire is to bring her or him back to the cinema time and again, to crave an unattainable, flickering fantasy life. Monroe became the brightest star, and one of the briefest. The star system, based crucially on idealized images of women, constituted these images as commodities which would, in a self-perpetuating cycle, generate further relations of exchange and increased profitability. Representations of women thus became the commodities that movie producers were able to exchange in return for money. For a while, Monroe was the producers' best friend.

Monroe was forced to depend for her security on the 'goodwill' and recognition of men – even to be interpreted by them in writing because, at least early on, she feared that sexual competition made women dislike her. Monroe once said, 'I don't mind living in a man's world as long as I can be a woman in it.' She was certainly not the passive blonde some writers make her seem, although men are typically at their most sceptical when considering characteristics (such as intellect) which contradict their fantasy image of woman. Probably more than any other female star of her era, Monroe made a public stand against such issues as nuclear weaponry, racial discrimination, anti-communism, and the star system itself. It was not that Monroe was separate from other women, but that they were separated from her. 'She was not a loner,' Joe Mankiewicz observed. 'She was just plain alone.' Even if they had wanted to (and many *did*), the women in her life lacked the power to protect her, or *nurture* her – her mother was no mother, her father was nowhere. 'Marilyn' was a male fantasy, fiercely protected by men from any threat – even from Monroe. In movies, photographs and prose, Monroe was mainly seen through male eyes. Mailer (1973: 15) says that 'Marilyn suggested sex might be difficult and dangerous with others, but ice cream with her.' Mailer's 'white Marilyn' is his self-made escape from the threat posed him by *real* female sexuality: he wants wonderful sex uncomplicated by anxiety about satisfying the woman. This male fantasy 'Marilyn' is made for men's pleasure: she giggles, she shakes her shoulders, she moves to his rhythm, she attends to his needs, she lives only when and where he wants her. She is far away from being Marilyn Monroe, the person men rewrote, the person I, inevitably, rewrite. Monroe actually disliked her name 'Marilyn' for several years after she was asked to adopt it; she never really enjoyed the movie 'Marilyns' she was obliged to portray. She loathed being seen as a mere sex symbol ('I thought symbols were something you clash') yet she was internationally regarded as the most potent embodiment of sexuality the screen ever saw, a sexuality which indeed seemed to emanate from her *own* exultation and fascination with her physique and her sensuality.

'I'm not interested in money,' she once told a producer, 'I just want to be wonderful.' More than wanting to be loved by you, Monroe was committed to perfecting her art, the art in her life and the life in her art, she wanted life to be wonderful. She wanted knowledge, she wanted physical power and beauty, she wanted children, she wanted (she needed) self-respect. The desire to be wonderful brings with it several supreme sacrifices. Even at the height of her success, there was a real anxiety about her, a deep desire to succeed *and* to please that animated her movie performances and was often quite moving, invoking in both women and men a powerful urge to offer her support. Like Brando and Dean, Monroe managed, with little apparent effort, to solicit pity from the audience. Billy Wilder, who directed her near the beginning and the end of her most memorable period, remarked: 'she made her name playing trivia, absolute, total trivia, and to make it playing trivia is much tougher than playing Ibsen.' The respect for Monroe is evident although even Wilder here speaks in a rather condescending tone of the 'plucky little lady' school: Wilder co-wrote two pieces of this 'absolute total trivia' (*The Seven Year Itch* and *Some Like It Hot*), and Monroe played a major role in making Wilder's movies highly respected as quality works.

One can find an alarming curve of cynicism moving from the sweaty, seedy, sexist prose of Norman Mailer to the supposedly 'sensitive' study by Gloria Steinem – a curve which seems to chart a growing anxiety concerning whether Monroe (between you and me) was really, after all, something of a 'dumb blonde'. Mailer's noisy posturings are, of course, obvious to the point of self-parody; his book is at least honest in its open obsession with Norman's panting pursuit of his very own fantasy 'Marilyn'. Steinem, famously feminist, is a more surprising cynic. She arrives on the biographical scene like a caseworker, retrieving Monroe's life from a faded manila folder. Her study is profoundly repressed in complexion, offering feminist uplift whilst an obsession with martyrdom bespeaks futility. It is vital to avoid conflating Steinem's study with the feminist study of Monroe: the weakness in the former is not a product of the latter. Monroe becomes a 'lesson' for Steinem to draw upon: indeed, as illegitimately as did her 'enemies', Steinem *uses* Monroe in what is ultimately an unfortunate manner. Mailer famously described Monroe as 'every man's love affair with America'. This affair became a burden for many when they were called upon to give as well as take. The screen 'Marilyn' was fun, the real Monroe was trouble. It seems that Monroe was expected to appear and disappear just as painlessly and as magically as her movie image; when she persisted, it seems not so many liked it hot. Henry Hathaway, who had directed Monroe in *Niagara,* argued: 'You don't have to hold an inquest to find out who killed Marilyn Monroe. Those bastards in the big executive chairs killed her.'

In a society where to be first was paramount – to get there fast was more vital than serving the future – the notion of built-in obsolescence took root: it had to be newer, neater, better, brighter, franchised, marketed as widely as possible, and then replaced. A common ironic observation regarding celebrity culture is that the one thing worse than being 'dead, but not forgotten' is being 'forgotten, but not dead'. The rapid rate of turnover in society applies both to artefacts and to people: Monroe was a star – what is brighter than a star, and what burns out more rapidly? Fame is America's prime export, recreating in materialistic terms the promise of perfectability which first drew Europeans to the New World. While the famous aspire to be the perfect product, attractive and addictive, the public craves fresh connections and new items on the menu. As a commodity, Marilyn Monroe was marketed as a New Name, a New Face, a New Star; she was used up, worn out and replaced. It was Monroe's tragedy, in such a heartless system, to be so much more than a mere commodity.

It is impossible, despite biographers' efforts, to determine what Monroe was feeling or hoping for during those final few days of her life. Her brief recorded comments are necessarily weak indicators: an optimistic remark may be said with heavy irony, a desire to commit suicide may be expressed lightly and half-heartedly. Piercing together the current collection of fragmentary facts, one may plausibly picture a woman who has suffered but who is strong enough to persevere, who can certainly cause hurt in others but who seems unusually incapable of consciously hurting others, spending days in dreams of distant hopes, visiting friends, talking business, crying, laughing, relaxing with old lovers, missing recent lovers, reading about herself, feeling angry, feeling hurt, feeling *wronged*. Perhaps all one can do is to gently brush one's knowledge against the skin of someone else's secrets, sensing a reality lurking in the silence of memory. Perhaps only by admitting the author to the library will one find any body: one acquires that rare passion, the passion for the other, the passion for that other that is within oneself, within myself. As Barthes (1979: 101) writes, 'I myself cannot (as an enamoured subject) construct my love story to the end: I am its poet (its bard) only for the beginning; the end, like my own death, belongs to others; it is up to them to write the fiction, the external, mythic narrative.' Reflecting on her own convictions and the conviction of her critics, Monroe once said: 'Goethe says a career is developed in public but talent is developed in silence. It's true for the actor. To really say what's in my heart, I'd rather show than to say. Even though I want people to understand, I'd much rather they understand on the screen' (in Goode 1986: 199).

'I don't think I've ever met a writer I'd like as my judge.' What am I doing to Monroe? I favour what Stanley Cavell describes as *empathic projection* – not merely an identification *of* Monroe but *with* her, as a human being. John Dunn (1985: 153) makes an important point when he writes:

> The claim to know better is flourished menacingly at identities, personal, cultural, and political, from the outside as much as it has ever been before in human history. But today . . . we know that it can be vindicated only within identities, that the only authority which it can possess is a human authority, an authority *for* human beings not an external domination over them.

Few writers have shown any desire to avoid dominating the memory of Monroe; Mailer, in particular, will box clever and box dirty to protect the 'Marilyn' he so skilfully boxed into his corner.

I feel most drawn to the image of Monroe which appears on the cover of this book. It is one from a series of photographs taken by Bert Stern just weeks before she died. The session produced some extraordinary pictures, with Stern organizing scenes that went perilously close to cliche, and Monroe investing tremendous energy in her repeated attempts to retain some originality. According to Stern (1982: 89), the image I selected was his most subtle: 'Marilyn put on the simplest black dress. . . She sat on a chair. She was beautiful. All I had to do now was backlight it. That image was of the essence of black and white . . . and blonde.' This singular image of Monroe caught my attention when I began writing this book and has remained in my mind. When I try to describe this feeling I court controversy: feminists have often pointed out that the enumeration of physical characteristics is always a kind of objectification – yet as I try to pinpoint the essence of the image, this is what happens. This Monroe *holds* me, though I cannot say why or *where:* is it the heavy eyelids, the smoothly curved neck, the soft-lit skin, the position of the hands, the slope of the shoulders? The effect is certain but unlocatable, ineffable, it does not find its name: it cries out in silence. By giving me the absolute past of the pose, the image shows me death in the future. In the face of this photograph of Monroe, it occurs to me: she is going to die within weeks of its execution. There is, as always, a 'defeat of time' in the image: *that* has gone, *that* is going to leave us. When Monroe was given a script about the Jean Harlow story, she sighed: 'I hope they don't do that to me after I'm gone.' She has gone, long gone, and they *have* done that to her, over and over again. I can understand why they did it – I cannot entirely escape doing it myself. In this image before me, the figure seems caught in history as in quotation marks, a grainy monograph, a visible wound which causes the memory to bleed. Knowing the pictorial literature on Monroe, I would have recognized Monroe among thousands of other women, yet I could not 'find' her: I recognized her differentially, not essentially. The image on the cover certainly touches me, captivates me in a curious way, yet I do not feel it leads me to Marilyn Monroe – not quite.

On the set of her final finished film, *The Misfits,* the photographer Henri Cartier-Bresson observed Monroe and studied her appeal: 'there's something extremely alert and vivid in her, an intelligence. It's her personality, it's a glance, it's something very tenuous, very vivid that disappears quickly, that appears again. You see it's all these elements of her beauty and also intelligence that makes the actress not only a model but a real woman expressing herself.' (in Goode 1986: 101). Throughout this book I have experienced this fascination, this frustration, the feeling that, just as my montage of images is set to capture the spirit of Monroe, she somehow eludes me. The attractive, exciting experience is of the *process* whereby 'Monroe' comes to resemble my 'Marilyn', momentarily making *me* feel I can know her. Buchner (1974: 87) has a description that seems to illustrate this point:

> When I walked up the valley yesterday, I saw two girls sitting on a
> rock: one was binding up her hair, the other helped her; and the
> golden hair hung down, and a serious pale face, and yet so young, and
> the black dress, and the other so eager to help. . . One might wish at
> times to be a Medusa's head to change such a group into stone so
> people can see it. They got up, the beautiful group was destroyed; but
> as they so descended between the rocks, it was yet another picture.
> The most beautiful pictures, the most swelling tones regroup, dissolve.
> Only one thing remains: an infinite beauty, which passes from one
> form to another.

'You try to photograph the truth', wrote Bert Stern (1982: 89), 'you frame it . . . you outline it using nothing but white light. Then you press the button. And if you're in tune, on that rare wavelength, you'll get to see the truth in black and white and circle it with red grease pencil.' Monroe experienced the positive and negative effects of being photogenic: since her early sessions as a model, she acquired a taste for posing and holding poses; everything was, at times, an opportunity to freeze a gesture, to instill a mood; in these brief, beautiful moments of timelessness Monroe became her own statue. Barthes (1979: 194) remarks that, 'the nature of the photograph is not to represent but to memorialize.' Monroe constitutes herself in the process of posing, making another body for herself, transforming herself in advance into an image, working on her skin from within. She poses, she knows she is posing, she wants you to know that she is posing, but this additional message must not alter the precious essence of her individuality: what she is. 'I wanted something no one else could get from Marilyn, captured in a photograph forever.' Bert Stern here expresses the desire of a generation of photographers, the request from a whole audience. The figure caught within the grain of the image seems so vulnerable, so unaware of what we are aware about her, of the times in store for her – the next moment, the next month. We could know this about ourselves, if we could move the force of nostalgia toward the anticipation of the fact that each moment is always stopped from every other. The still Monroe has a calmness, in her somewhat languid way of lowering her eyelids, of stretching her neck forward, that promises something hidden, as her smile seems to hide behind the very act of smiling. She is shown in every attitude, every angle, every guise, an identity fragmented into a powder of images. The photographer pursues life as it leaves, a hunter of the unattainable. There were many possible photographs of Monroe, and many Monroes impossible to photograph, and perhaps what we seek is the unique photograph that will contain the former and the latter – an impossible, yet irresistible, compulsion. To keep Monroe within the range of hidden lenses, to 'snap' her not only without letting oneself be seen but without seeing her, to surprise her as she was in the absence of one's gaze, of any gaze: it is an invisible Monroe we want to possess, a Monroe alone, a Monroe whose presence presupposes the absence of me, you and everyone else.

'Some people have been unkind,' Monroe complained. 'If I say I want to grow as an actress, they look at my figure. If I say I want to develop, to learn my craft, they laugh. Somehow they don't expect me to be serious about my work. I'm more serious about that than anything, (in Rollyson 1986: 121). The tension between Monroe's public and private selves, her professional and personal lives, is particularly evident in her image as movie star. The movies demanded that she seem dumb, central and giving; her society demanded that she seem rational, marginal and grateful. Monroe was isolated for others and from others. Hollywood deploys love within a framework of melodrama so that it is seen to be neurotic, emotional and instinctive (which it may be), and not rational, intellectual and personal (which it can also be). The audience was attracted and entangled within the illusion of love (the *projection* of love), leaving them in love with screenplay love, rather than attached to other people, other lovers, other loves. Looking at Monroe's movies today, with a view towards 'seeing Marilyn', we face the fundamental qualities of cinema. To wish to view Monroe herself is to wish, to work, for the condition of viewing as such. Our modern condition has become one in which our natural mode of perception is to view, feeling unseen, unheard, in suspense. We peer out at the world from behind the self: it is our fantasies, now all but thwarted and out of hand, which are unseen. Movies persuade us of the world's reality in the only way we have to be persuaded, without learning to draw the world closer to the heart's desire: by taking views of it. Monroe once wrote the lines, 'Life – / I am of both your directions' (in Rosten 1980: 57). As one reel winds on the other runs down, the movie does not project the 'real Marilyn', but it may well present the poignancy of her absence.

In her last few weeks alive, Monroe began to narrate her life story to journalist George Barris; the project was intended as a way of 'refuting' the studio image of the 'dumb blonde' who had become dizzy and difficult. The project died with her, and her life became a cultural battlefield upon which the biographers have fought for the territory of her reality, the solid essence beneath the ethereal enigma. Ironically, the more complex her image becomes, the easier it is for that image to be appropriated. Sartre's (1976) fictional biographer reflected, 'Nothing happens when you live. The scenery changes, people come in and go out, that's all. There are no beginnings. . . But everything changes when you tell about life; it's a change no one notices; the proof is that people talk about true stories.' Increasingly, the traditional practice of biographers – the chronological and comprehensive life – is incommensurate with what we know about the complexity of individual lives. Today, new demands are placed upon biography from psychology, anthropology, sociology, and history; as a literary enterprise, biography must respond by registering in its form and content new means of expressing human experience. Dissatisfaction with previous biographies of an individual seems endemic to readers and writers alike. Consequently, multiple lives of major figures are not uncommon: the extraordinary library featuring Monroe is confirmation of this fact. Why should we need so many?

As a genre, biography continually unsettles the past; versions of a life are necessary moments in the development of the form as well as in the understanding of the subject. No biographer merely records a life. How the biographer expresses the life becomes, to some extent, the real subject of the biography. The men who have written about Marilyn Monroe have withheld themselves from genuine emotional involvement, and to that degree are unwilling to *be* the woman. Rather than opening themselves up, they try to open *her* up. The unfortunate process is echoed by lines from a poem Monroe used to carry around with her, Oscar Wilde's *The Ballad of Reading Gaol*:

And all men kill the thing they love –
By all let this be heard.
Some do it with a bitter look,
Some with a flattering word.

Embracing this problem, I made a sincere effort to include an awareness of my own involvement in this study, my obligation to observe my own inevitable prejudices. Thus, my aim has been to preserve some notion of identity, some appreciation that my subject was not simply a 'social construction' but a real person, an individual, a woman, with hopes and fears and strengths and weaknesses – some of which are alien to me, others which are very familiar indeed. In other words, I felt an obligation not merely to theatricality but mainly to *authenticity,* to the presence of the person – whether she be in the library or between the lines or in the mind.

The ambiguity of myth is a problem: on the one hand it suggests the essence of a person and, on the other, the legend that person has created. For biography this is especially problematic, for it finds itself with a dual activity, one assigned, the other assumed. The first is the desire to correct or revise the myth; the second is its own unconscious creation of new myths. No biography can duplicate the life of Monroe; despite all claims to objectivity, we read a *personal* view of her, not a documentary history. Myth emerges out of the author's need and the reader's desire for wholeness and order. Irony emerges out of the tension between the impulse to correct in the biography and its generation of new myths about the subject. Biography necessarily universalizes the more it individualizes, as it reveals the common experience of such conditions as triumph, love or failure. The myth of Monroe is ambiguous: is she 'Norma Jeane' or 'Marilyn', Monroe or 'MM', one or the other or both? Sociology is more comfortable when analysing cultural generalities than when appreciating human individualities. Monroe's studio tried to type-cast her, and now her biographers try again.

At a loss for what to do with the person who was Marilyn Monroe, critics opt for insult by caricature (the victim, the vamp). It appears at times as though every 'investigative journalist' is saying or writing dreadful things about Monroe: so dreadful, in fact, that they are unwilling to account for their views beyond the casual quip or the flippant rejection. This myth does not deny history so much as banalize it: 'I envision myself nibbled up by others' words, dissolved in the ether of Gossip. And the Gossip will continue without my constituting any further part of it, no longer its object: a linguistic energy, trivial and tireless, will triumph over my very memory' (Barthes 1979: 196). The figure is possessed by the writing like a piece of glass by light: pouting, suggestive, submissive, subversive, erotic, available, forbidden, sometimes naked, sometimes overdressed, and always blonde. A woman 'half-way between myth and reality', framed in the radioactive years of fifties America, the time of McCarthy and hydrogen explosions, the era of the anatomic bombshell. 'I am searching the other's body, as if I wanted to see what was inside it, as if the mechanical cause of my desire were in the adverse body (I am like those children who take a clock apart in order to find out what time is)' (Barthes 1979: 71).

Pulled apart, pulled together, pulled into place, the myth of Monroe is positioned for perusal. Biographers chart the succession of orphanages; the severed emotional roots; the series of producers, photographers, patrons and agents; the make-up and masseurs for body and mind; the marriages to two wealthy cohorts – Joe DiMaggio of Yankee Stadium and Fisherman's Wharf, then Arthur Miller of Brooklyn Heights and Reform Judaism; the curious connection with the Kennedys. From the crowd to the crucible to the corpse, and in between some thirty screen appearances, thousands of photographs, millions of words. The person, as Foucault says, is thus held 'under the gaze of a permanent corpus of knowledge'. Monroe's life, seen as a 'life-story', is a poem, and the problem of each biography is thus a problem of translation.

'Allegories are, in the realm of thoughts, what ruins are in the realm of things.' Benjamin's observation reverberates through the stream of words that threatens to wash and wear away the stains on the social conscience; as Barthes (1979: 181) has said, 'In the lover's very tears, our society represses its own timelessness, thereby turning the weeping lover into a lost object whose repression is necessary to its "health".' We can say that when a myth can no longer support revision – the being looked-over again – then the myth has died, we have died to it. So strong was the feeling when Monroe's death was announced that many people had 'rescue fantasies' about her: if only I had said this, done that, gone there, loved her in a certain way. . . Men thought that their protection might have saved her, women wondered if their empathy could have helped her. What one mourns for those who have died is the loss of their hopes. It is a measure of Monroe's vivid status as an American symbol that she is still written about as though it is not too late to bring her back to life. Mailer's *Marilyn* was the long, brawling affair he never had with the star, and Steinem's *Marilyn* made a rescue bid a quarter-century *after* the subject's death. Necessarily, there must be an attempt to develop an imaginary relationship between biographer and subject, a continuous dialogue between the two as they move over the same historical ground. It is imaginary, for obviously the subject cannot really, literally, respond; but the biographer must come to act and think and feel for the subject as deeply as possible. Monroe's identity as a professional performer made her in a sense, her own interpreter – sometimes a very frustrated one. Acting encouraged her oscillating notion of identity: her private self and her screen self inflected each other, sometimes sticking together, sometimes causing sparks to fly. In her last few months she was said to have withdrawn some evenings into her room, listening over and over to the Frank Sinatra song, 'Every Day I Have the Blues', perhaps reflecting on her own sense of self and identity when the line was sung: 'It's you I hate to lose'.

Monroe's fame frustrated her need, as an ambitious actor, to observe other people: celebrity breeds self-consciousness, it *aggravates* the self so much that one loses that ability to appraise, evaluate and record other people. Monroe came to see that 'Marilyn' was an intrinsic part of her, that she could never go back to being Norma Jeane, to being 'normal'. Her occasional expressions of anger and frustration were fuelled by a feeling that, in some sense, she had contributed to her own confused condition. In Hollywood she was a tough competitor, proud and aggressive when she liked, and she worked to develop her singularity *and* her popularity – a thankless task in a system demanding conformity to the standardized rituals of success. Invited to become her own product, she chose instead to produce her own personality. After her miscarriage she said, 'What good is it being Marilyn Monroe? Why can't I just be an ordinary woman? A woman who can have a family.' Drawing on her strengths, her skill, she resolved: 'If I can't be a mother, I'd better be an actress. I have to be something. And, whatever it is, I'm going to be good at it!'

The fascination of Monroe's story lies in the passionate sureness with which not only she but everyone around her played their parts, fusing and confusing life and art. How apt that all she said in her first recorded film should have ended on the cutting-room floor, bar the one word 'Hello'. How fitting the real debut in *The Asphalt Jungle*, her form glimpsed rising from the shadows like Nijinsky's faun. Her biographers have perceived and perpetuated this fearful symmetry. Monroe was a child of Los Angeles as well as a star of Hollywood; she was born into a highly industrialized, urbanized milieu – the principal place wherein artifice, *acting,* was a recognized occupation, a lifestyle in which the authentic was hardest to discern. The dictum from Henry James – 'Never say you know the last word about any human heart' – is never more apposite than when considering Marilyn Monroe. A writer who knew Monroe and who wrote an invaluable memoir of the experience, W. J. Weatherby, has commented to me: 'It needs a really fine biographer with no axe to grind and able to see Marilyn as she was beneath all the sexual camouflage, and he/she will probably not come along until we have more distance from her and her contemporaries.'

Biography remains an assemblage of abstract contents unless it is given a form that is derived from insights into the personality of the subject herself. The autonomy of Marilyn Monroe is playing against a history. Reading the text and looking at the pictures of Monroe, it is terribly easy to forget that she is no longer here, somewhere (there is a place for her). As Richard Rorty (1987) has argued, there is a very disturbing irony in the pleasure of redescribing a figure and the suffering of the figure being redescribed. Monroe was someone intent on being described in her own terms, despite working in a society intent on redescribing women in men's terms. Monroe was redescribed as Arthur Miller's 'Marilyn', Hollywood's 'Marilyn', and the critics' 'Marilyn'; today she is redescribed as Mailer's 'Marilyn' or the *MS* 'Marilyn'; none of these versions respect Monroe's wish, and in a sense they *cannot* do as she wished, but they *could* have tried. In redescription, there is a constant danger of building an image upon insensitivity; what was important for Monroe becomes worthless for her biographer, what was precious is now discardable, what was valuable is now valueless. My Monroe is necessarily a Monroe redescribed, *reconstituted;* part of the Monroe you have been reading is my Monroe, is myself, and my only hope of respecting her wishes has been to draw attention to the ways in which our bodies betray our isolation from each other. Positively, my approach has been concerned to highlight those rare opportunities whereby we can move each other, care for the other, to imaginatively touch their point of view.

Monroe always opposed the movie industry's theatricality with a strong sense of authenticity: 'Fame has a special burden. . . I don't mind being burdened with being glamorous and sexual. But what goes with it can be a burden. . . I feel that beauty and femininity are ageless and can't be contrived, and glamor – although the manufacturers won't like this – cannot be manufactured. Not real glamor, it's based on femininity' (in Meryman 1962). This sense of authenticity would appear to offer an attractive line of communication between the biographer and the subject, providing not an unusually 'pure' sighting of the person but at least an exceptionally *persuasive* one. Unfortunately, few biographers have felt inclined to listen to Monroe, preferring to 'write over' her, talk over her:

> So we think of Marilyn who was every man's love affair with America, Marilyn Monroe who was blonde and beautiful and had a sweet little rinky-dink of a voice and all the cleanliness of all the clean American backyards. She was our angel, the sweet angel of sex, and the sugar of sex came up from her like a resonance of sound in the clearest grain of a violin. (Mailer 1973: 15)

Norman Mailer punched out these lines in 1973. He proceeded to describe Monroe as 'a very Stradivarius of sex': one wonders if he had ever come across Monroe's own remark: 'An actor is supposed to be a sensitive instrument. Isaac Stern takes good care of his violin. What if everybody jumped on his violin?' We are back full circle: 'I hope they don't do that to *me* after I'm gone.'

'An actor is supposed to be. . .' I wish could have *heard* Monroe say that and know in what context it was conceived and spoken. Did she say it mischievously, tentatively or angrily, did she mean it seriously, humorously or as a sad statement of fact? Any or all of those interpretations are appropriate; she said it the year she died. We are forced to come to terms with the anxieties of interpretation, the attempt at empathy. Monroe's words seem now spoken amidst her own silence: language assumes its most intriguing form at those moments of attempted closeness between two bodies. Intimacy breeds irony: in diagnosis ('tell me where it hurts'), in lovemaking *('No!/Yes!')*. Barthes, in *A Lover's Discourse* (1979), observes that, at the centre of our lives, there is an area of experience where the rules of common sense and the norms of narrative collapse: the realm of *compassionate sentiment*. In the amorous moments of our lives, we experience intensely yet find that the language on which we normally rely cannot be employed to ascribe sense to this experience. Each time we mean to utter a great truth we hear ourselves speaking banalities: *'I love you more than words can say'; 'Where have you been all my life?'*

Monroe certainly seems to speak directly to us – that is part of her appeal, as she appreciated: 'somehow I feel they know that I mean what I do – both when I'm acting on the screen or when if I see them in person and greet them – that I really always do mean hello and how are you?' (in Meryman 1962). It is striking how strongly Monroe's own voice seems to pierce through the impersonality of the printed page. 'Seems' is the necessarily qualified term to use, for the body in question is now most evidently the material body of the book: 'My bones are leather and cardboard, my parchment flesh smells of glue and mildew. . . I'm taken, opened, spread out on a table, opened with the flat of the hand and sometimes made to crack' (Sartre 1964: 161) This reconstituting, rebinding of the body makes me read one of Monroe's own lyric fragments in this nostalgic light:

> I stood beneath your limbs
> and you flowered and finally clung to me
> and when the wind struck with . . . the earth
> and sand – you clung to me.
> (In Rosten 1980: 59)

Billy Wilder, whose relationship with Monroe was often a stormy one, now confesses: 'We just happen to miss her like hell!. . . [N]ever a week passes when I don't wish she was still around.' How does the biographer express this sentiment? As Barthes (1986: 296–305) wrote (in a text left all but complete at the time of his death), one inevitably fails at trying to talk about what one loves. Perhaps the problem is more a want of conviction rather than a want of words: if the story of Marilyn Monroe reads at times like a nightmare, then it is precisely because, at times, it *was* a nightmare. Beneath and beyond the biographical form, there is the story of the abject neglect of a sensitive and exploited human being whom no one was finally willing to help in any genuinely positive, selfless way. She was 'one of the most loved public figures of the century', and also one of the most ill-treated. The inhumanity of this story is almost unfathomable, at times unbearable. The depiction of Monroe as a figure not deserving this seriousness is an unforgivable act by those shielded behind scepticism; as Adorno (1978: 112) argues, 'the scentless bouquet, the institutionalized remembrance, kills what still lingers by the very act of preserving it. The fleeting moment can live in the murmur of forgetfulness, that the ray will one day touch to brightness; the moment we want to possess is lost already.' The possessiveness is pervasive: through translation ('So Marilyn meant *this* . . .'), through review and synopsis ('So what *you* are saying is *this*. . .'), and through reproduction (the image, the revival, the memento). W. J. Weatherby treated Monroe with respect and found her responding with warmth; he remembers her telling him (1976: 174), 'You're a good listener. You don't want anything from me, which is rare.' An American critic, Bill Stout, has said: 'digging into old stories of death, coming up with theories, making accusations that cannot be answered . . . well, that kind of stuff makes money.' If Monroe's memory *is* being exploited, it is surely better to reclaim it rather than to renounce it.

How can one, how should one, 'reclaim' Marilyn Monroe? Layer upon layer, lover upon lover, the laminations become prodigious, promiscuous, incestuous; each image adds to the labyrinth of myth, a labyrinth which never *leads* anywhere. The process of relating oneself to the other holds a fascination of its own: 'lovers incessantly "drink in speech upon the lips of the beloved", etc. What they then delight in is, in the erotic encounter that play of meaning which opens and breaks off: the function *which is disturbed:* in a word: *the stammered body*' (Barthes 1977: 141). The more I read of the writings on Monroe, the less I feel that I know her, and the more I want to try to know about her. Yet, that shadowy figure on the cover of this book seems to grow visibly less tolerant with each stroke from my pen. As I struggle to express the propriety of my interest, I draw nearer to the recognition, and the practice of tautology: *the authentic is what is authentic,* I am fascinated and '*I am fascinated*' – two moments in my meditation, hovering over the subject like two guardian angels. Why are we fascinated? If ever there was a poorly treated figure of 'society', Monroe was that figure – of a society that professes dedication to the relief of suffering whilst degrading the joyous. She was no helpless victim, but she *was* often harmed. The evil of a cultural climate is cultivated by those who share it. Anyone who has ever entertained resentment against the good for being good, against the vulnerable for being vulnerable, is the opponent of Monroe. She actively sought humanization, and emotionally she fought against becoming a symbol or an object; to be sexually attractive was one thing, but to become a sexual fetish was another. Monroe was not just haunted by 'Norma Jeane', she was also haunted by '38-24-36'.

Milton Shulman once wrote: 'Whenever a man thinks of Marilyn Monroe, he smiles at his own thoughts' (in Taylor 1984). The thought is sometimes a sinister one. Too many literary abstractions which present themselves as universal have in fact described only male perceptions, experiences and options, and have misrepresented the social and personal contexts in which writing is produced and consumed. Furthermore, the exploitation of the female audience, common in popular culture, is especially relevant to the case of Marilyn Monroe. Women are drawn to the woman who is witty, talented, dignified, sometimes vulnerable, sometimes strong; they are disenchanted with the image drawn by men, taken by men, marketed by men – the 'dumb blonde', the 'sex goddess', the 'sweet angel of sex'. In order to reclaim Monroe as a person, to make her memory open for other women, the standard representation must be deconstructed – and the men who manufactured that representation must examine and revise their own image and identity. *Some Like It Hot* ended with a man saying of men: 'Nobody's perfect!' Now men must make this go 'beyond a joke'. As Stephen Heath writes in 'Male Feminism' (1984: 270):

> no matter how "sincere", "sympathetic" or whatever, we are always in a male position which brings with it all the implications of domination and appropriation, everything precisely that is being challenged, that has to be altered. . . Men are the objects [of feminism], part of the analysis, agents of the structure to be transformed, representation in, carriers of the patriarchal mode.

Men *can* do something to help, contribute and criticize, but they must resist the temptation to assume women's space, to colonize, to muscle in and make feminism *theirs*. This temptation, if succumbed to, draws one's self-analysis to a premature close, banalizing one's argument and touching the dishonest in a manner noted by Wittgenstein (1980: 8):

'When you bump against the limits of your own honesty it is as though your thoughts get into a whirlpool, an infinite regress: You can say what you like, it takes you no further.' Feminism unsettles men's assumed positions, undermines given identities. Men can make themselves receptive to feminism, they can listen and learn and try to *respond* to feminism. In a very long tradition, sexual fascination has been an important motive to philosophy, as if to acknowledge and appreciate the intimacy and mutuality of one person's investigation of another. The concern, the passion, the admiration that philosophy can teach is the power to accept intimacy without taking it personally. As women work to redefine the site of sexuality, I try to learn from them: this learning I work at with a mixture of anxiety, pleasure and resistance. What I have attempted to do with the myth of Monroe is to cancel out the other male texts, to consciously counteract the distortions fostered by men's uncritical styles. Monroe, for these men, is a mystery to be 'looked into' by the inquisitive male investigator: he wants to get things straight, to know who she is and who they are not. At the same time as men write Monroe's womanhood, they also write their manhood. The sense that all one really knows is one's own experience should be admitted when speaking of any experience so intimate as the sense of one's own sex.

Interpreting Monroe, trying to think her thoughts, I felt both pleasure and painful self-doubt. Monroe wrote that her critics were 'white-masking' themselves by calling her 'the lewd one'; I sometimes wondered whether I, against my own convictions, was not white-masking *myself* by focusing upon Mailer and his colleagues. The thought caused a tension in my writing: 'The linings of my heart keep sticking together and to open it I should each time have to tear them apart' (Wittgenstein 1980: 57). 'If only', said Monroe, 'they would be honest – just once. . .' There is a part of me dedicated to avoiding any admission of fault or guilt. I am also prone to the desire to possess: whilst I recognize the worth in wishing the loved object 'just to be', within me there is the wish for her to be for *me*. The temptation that I am most susceptible to is the use of feminism as a (re)assurance of an acceptable identity, right and proper, a personal and political position whereby she will find me attractive. I *hope* I have succeeded in producing a feminist criticism which (usually) avoids feminist impersonation.

Flaubert's Bouvard and Pecuchet manage to move from library to library whilst remaining thoroughly unmoved by the knowledge: as Monroe said of her interpreters, 'they were never there.' I wanted to draw attention to the role of the writer in the representation of the woman, showing the sense in which the body in the library is my body, with stacks full of memories, a catalogue of personal references. Rare and memorable moments are filed, in which we become aware of the complexity and degree of skill required to meet demands made upon us: when we fall in love, when we enter a new environment, when we feel it necessary to deceive those around us, when our lives are disrupted by some event that shows us that others have been seeing us in ways of which we were quite unaware, and when we find that love breaks down. In these key moments, we appreciate both the nature of the strategic tasks in which we are involved and the opacity of the conduct of those others around us. Thus, our moments of self-knowledge coincide with the moments when we acknowledge our lack of knowledge of others. We begin producing by reproducing the person we *want* to be. When I attempt to examine my interest in Marilyn Monroe (not selfishly but sincerely), I am necessarily involved in an effort to examine and defend my interest in my *own* experience, in my own body, in the moment and passages of my life I have spent with (and without) my subject. I am thus moved to defend the process of criticism. A critic who merely records previous interpretations in an impersonal, 'objective' way, is no real critic; as Tanner (1987) rightly says, 'To sow your book with dubiety is a too easy way of insuring yourself against criticisms. As is the attempt to pre-empt or defuse criticism by anticipating it and writing it into the book.' A work one cares about is not so much something one has read as something one is a reader of; connection with it continues and deepens, as with any relation one cares about. 'In everyone there sleeps a sense of life according to love' (Philip Larkin). Once we acknowledge that, in a sense, the objects have chosen us, have stood out to us, caught our eye, and demanded our response, then our interest goes beyond ourselves. If we are prepared to live at all decently, we can *trust,* amid our doubts, that our choices of what to value and how to express our values are not indifferent. Interpretation is thus a form of participation, an involvement.

Looking at Marilyn Monroe involves looking at the power of celebrity and the extension of its reach in recent decades. The issues presented by celebrity power finally merge with still larger cultural issues and cannot be isolated from them. Making a myth of accomplishment, celebrities become pseudo-events, diverting the public gaze from the real to the ideal. The machinery of celebrity labours to produce a sense of the culture's greatness, a star to assuage society's fear of decline. Nobody rests in celebrity culture: the star's pursuit of immortality involves the exploration, testing and extension of power; the consumer craves the 'newness' in the well-known, the freshness of the famous. When Saul Bellow was awarded the Nobel Prize, he wondered aloud if his 'dream space' might now be swallowed up by his celebrity. Marilyn Monroe suffered from the erosion of this space: her creativity involved contemplation, her celebrity involved presentation. She had to provide intimacy without privacy. How much of her hold on us is due to the star machine's wheels and pulleys, how much is due to her personal qualities? Certainly many movie stars have moved us without receiving from us the kind of affection we bestow on Monroe. Celebrity, the most visible aspect of popular culture, enables the stars to escape the categories of their origins, to assume both a wider public life and to make a profound emotional impact on the individual's inner life. Monroe's public image is great and her impact immense, for she brought sincerity and warmth to the mass produced icon of the market-place, using her vulnerability as her private mode of communicating to us. As Rosten (1987: 107) recalls of Monroe's relationship with the public: 'They genuinely wanted to wish her well, they *liked* her.'

Monroe has said: 'you know when you get grown up and you can get kind of sour, I mean, that's the way it can go.' Her biographers boast an intimate knowledge of their subject, feeling her warmth, listening to her breathing, reading her mind. They generally provide a picture that seems the work of someone 'sour', sceptical, keen to favour cynicism for fear of sentimentality. Monroe exudes innocence, fragility and deprivation – this makes the 'sex-pot' stereotyping profoundly shaming. She came into conflict with a succession of men whose education had sensitized them to their own sex and left them ignorant of the other which is not to be known but conquered. It is pointless to avoid that often-said, often-cited fact: Monroe *is* in many ways a most tragic figure, but a tragic figure who is easily associated with the happy things in life. Arnold Newman, who consulted Monroe on her musical numbers, describes her as 'an exceptionally warm, compassionate, surprisingly self-conscious girl. I found her to be fiercely loyal – sometimes to a fault.' Monroe, it seems, was someone who knew how to make a gift of herself, yet did not know how to receive. There was never anyone to teach her how to respond to the often graceful tenderness of gifts; we all, at times, find it hard to comprehend how others can love us for our own sake. Monroe had to struggle to thoughtfully receive blessings, though her world can have offered her few. Any portrait of her shows the tension: look at the eyes, they are not happy, yet within them lies fortune as it lies in gambling or in love, a relentless, explosive will to happiness. This Monroe, as I now know her, is an image I can neither really remember nor entirely erase. At best she seems complex, fascinating, *real;* at worst she seems a bland, manufactured icon. The good critic, animating an array of facts, weaving a tissue of speculation into a coherent scene, can make us glimpse an authentic image. I am reminded of Monroe's comments: 'It's often just enough to *be* with someone. I don't even need to touch them. Not even talk. A feeling passes between you both. You're not alone.' There are fleeting moments in which it can seem, when one *trusts* the critic's argument, that you are me and I am you. In the most personal of insights, there is space for the most helpful generalities: embracing the other's knowledge, touching on some shared sensibility, the reader suddenly wonders: *how can they possibly know that I feel those feelings?*

Monroe's most memorable performances touch us, make us sit in the silence and darkness and think to ourselves: yes, I know exactly what she means, how she feels. Her very openness makes it particularly hard to resist assuming her position and speaking on her behalf. I like to perceive my response as in accordance with a sense of *admiration,* an admiration discussed by Irigaray in relation to an ethics of sexual difference: 'What has never existed between the sexes. Admiration keeping the two sexes unsubstitutable in the fact of their difference. Maintaining a free and engaging space between them, a possibility of separation and alliance' (in Heath 1984). She adds: 'I will never be in a man's place, a man will never be in mine. Whatever the possible identifications, one will never exactly occupy the place of the other – they are irrreducible the one to the other.' I have tried to respect this irreducibility, and used it as my starting point. One of my intentions has been to demonstrate that Marilyn Monroe can be written about with the same seriousness and respect that any important *male* artist receives and deserves.

The best passages in the best writing on Monroe exhibit an admiration for her and her work. The quality shimmers in the deep background with a hint of proximate revelation: the eye grows used to the darkness and the flicker of private meaning, one sees the full figure, part of the good, part of the bad. Barthes (1981: 25) has called that part of a photograph that creates a special current of personal pleasure, that catches our attention, the *punctum.* Within her movies, there are certain actions, exchanges, expressions, which seem inconsequential moments in the narrative. Yet these very moments work in me, after I leave the cinema, and they return – becoming memories, tenacious memories, making me happy and sad, retrospectively. What do we remember when we recall the loved object, what succeeds in touching the one who survives to remember? What do I remember? I remember the 'Diamonds are a Girl's Best Friend' number from *Gentlemen Prefer Blondes,* reds, pinks and yellows, a distinctive look of self-assured mischief. I remember 'the Girl' in *The Seven Year Itch* demonstrating her 'Dazzledent' toothpaste smile, an effortless performance which somehow seems to reinforce the magic of the expression rather than revealing its theatricality. I remember Monroe interviewed on Ed Murrow's *Person to Person* show, looking so young and nervous, each word spoken sounding like a precious confession, eyes looking out at the camera with a trust that knows betrayal. I remember one of the last scenes from *Bus Stop,* with Cherie giving her emotional speech, the camera moving in close, and as she moves her head away from the hand where it rested, a delicate string of saliva is shown. I remember Elsie Marina in the Grand Duke's rooms, almost dancing as she dodges the officious servants wheeling in the dinner trolleys. I remember Sugar Kane's extraordinary playfulness when she sneaks into Daphne's train berth, and her melancholic appearance as she sings of being 'through with love'. I remember the climactic scene from *The Misfits,* crouched, cornered, crying *'Liars!'.* Perhaps the most curious and most persistent memory I have is of an earlier scene from the same movie: Gay and Roslyn are at the breakfast table, the camera focuses upon Roslyn, wrapped up in a white dressing gown, the sunlight playing over her hair, she leans forward, pauses, gives a simple subtle smile, and says, 'You like me Hunh'?' The director, John Huston, is said to have been so moved by this very brief and innocent scene that he embraced Monroe and thanked her. It never fails to move me when I see it.

My remembrance of Monroe turns over scenes, sayings, stage after stage, reel after reel, figure after figure, phrase after phrase, further meanings found in each moment. We carry within us our own private cinema, screening our personal memories of Monroe, projecting flickering, fantastic images of our own 'Marilyn', for us, for now, for ever. Faced with the image of absence, the frivolous insignificance of language amidst such stillness seems the very space of love, its subtle music. As I write this, I would like to believe that the sorry, sordid stories, which spiral upwards and outwards will fail to detract from the underlying *humanity* of Marilyn Monroe. We are sentenced to seeing her in quotation marks, but we can try to prevent the marks from harming the memory. What I remember is not so much a being as a *quality:* the irreplaceable. As I look up, I glance once again at the fragile figure of the person who died only months after I was born: there she is framed, her hand curving around her cheek, her hair bound back, her body sheltered by darkness. After all the pictures and all the prose, there is silence, there she is and here am I.

'I guess,' said Monroe, 'I wanted love more than anything in the world.' So I guess, do I. My interest in Marilyn Monroe has been a recent one; for most of my life I knew little about her, I knew only what everyone knows about her. One thing that I had not anticipated when I began this book was how much I would come to miss her. Marilyn Monroe was a performer, a person on whom no one's fate depended, and yet her energy, sensitivity and openness to life made a profound impression on total strangers. Most movingly, there is her last interview: 'What I really want to say: that what the world really needs is a real feeling of kinship.

Everybody: stars, laborers, Negroes, Jews, Arabs. We are all brothers. . . Please don't make me a joke. End the interview with what I believe.' Hollywood did not kill Marilyn Monroe, neither did her critics, but they did fail to understand her and the pain and effort it took her to be what she was and to be as good as she was. It is, sadly, far too late to save Marilyn Monroe; perhaps, if we care enough for ourselves as well as for her, we still have time to do as she wished – to take her seriously. The concluding lines of this book were meant to express my admiration for my subject; with much pleasure, I found that the lines had already been written – by Monroe herself (1974: 55): 'I used to say to myself, what the devil have you got to be proud about Marilyn Monroe? And I'd answer, "Everything, everything."'

References

Adorno, T. W. (1941) 'On Popular Music' *SPSS* 9(1).

———— (1973) *Negative Dialectics* London, Routledge & Kegan Paul.

———— (1978) *Minima Moralia* London, New Left Books.

———— (1984) *Aesthetic Theory* London, Routledge & Kegan Paul.

Agan, P. (1979) *The Decline and Fall of the Love Goddesses* New York, Pinnacle.

Alanen, A. (1982) *Marilyn Monroe* Painatus, Valtion painatuskeskus.

Alberoni, F. (1962) 'L'élite irresponsable' *Ikon* 12–40(1).

Allen, W. (1980) *Getting Even* London, W. H. Allen.

Amengual, B. (1957) 'Marilyn chérie' *Cahiers du Cinéma* 73, July.

Anderson, J. (1983) *Marilyn Monroe* New York, Crescent.

Archer, R. and Simmonds, D. (1986) *A Star is Torn* London, Virago.

Axelrod, G. and Shaw, S. (1955) *Marilyn Monroe as the Girl: The Making of The Seven Year Itch* New York, Ballantine Books.

Babuscio, J. 'Marilyn on my mind' *Gay News* 85.

Banner, L. (1983) *American Beauty* New York, Alfred Knopf.

Barnes, J. (1984) *Flaubert's Parrot* London, Picador.

Barthes, R. (1975) *The Pleasure of the Text* New York, Hill and Wang.

———— (1977a) *Roland Barthes,* trs. Richard Howard, London, Macmillan.

———— (1977b) Interview in *Playboy* September.

———— (1979) *A Lover's Discourse,* trs. Richard Howard, London, Jonathan Cape.

———— (1981) *Camera Lucida,* trs. Richard Howard, London, Fontana.

———— (1986) *The Rustle of Language* Oxford, Basil Blackwell.

Beaton, C. (1957) *The Face of the World* New York, John Day.

de Beauvoir, S. (1984) *The Second Sex* Harmondsworth, Penguin.

Beckett, S. (1982) *Ill Seen Ill Said* London, John Calder.

———— (1983) 'Imagination Dead Imagine' *A Samuel Beckett Reader,* ed.

John Calder, London, Picador.

Bego, M. (1986) *The Best of Modern Screen* London, Columbus Books.

Bellow, S. (1975) *Humboldt's Gift* New York, Viking Press.

Belmont. G. et al. (1975) *Marilyn Monroe* Paris, Film Editions.

Benjamin, W. (1935) 'The Work of Art in the Age of Mechanical Reproduction' In Solomon, M. (1979) *Marxism and Art* Sussex, Harvester.

———— (1979) *One Way Street* London, New Left Books.

———— (1982) *Illuminations* Glasgow, Fontana.

———— (1985) *Charles Baudelaire* London, New Left Books.

Berger, J. (1984) 'The Secretary of Death' *The White Bird* London, Chatto & Windus.

Berkow, R. (1977) *The Merck Manual* New Jersey, Merck Sharp Dohme Research Laboratories.

Bernstein, W. (1973) 'Monroe's Last Picture Show' *Esquire* July.

Bessie, A. (1966) *The Symbol* London, The Bodley Head.

Boorstin, D. (1962) *The Image* London, Weidenfeld & Nicolson.

Bruce, W. (1954) 'Meet the New Marilyn Monroe' *Movieland* 12(11), November.

Büchner, G. (1974) *Samtliche Werke und Briefe* Munich, Carl Hanser.

Capell, F. A. (1969) *The Strange Death of Marilyn Monroe* New Jersey, Herald of Freedom Press.

Capote, T. (1981) *Music for Chameleons* London, Hamish Hamilton.

Carpozi, G. (1962) *The Agony of Marilyn Monroe* London, World Distributors.

Carrol, L. (1979) *Through the Looking Glass* London, J. M. Dent & Sons.

Cavell, S. (1979) *The World Viewed,* enlarged edn, Cambridge, Mass., Harvard University Press.

———— (1981) *Pursuits of Happiness* Cambridge, Mass., Harvard University Press.

Cixious, H. (1984) '12 Aout 1980' *Boundary* 2(12), summer.

Collier, P. and Horowitz, D. (1984) *The Kennedys* London, Secker & Warburg.

Colmar, M. (1979) *Whalebone to See-Through: A History of Body Packaging* London, Jackson & Bacon.

Condren, C. (1985) *The Status and Appraisal of Classic Texts* New Jersey, Princeton University Press.

Conover, D. (1981) *Finding Marilyn* New York, Grosset & Dunlap.

Conway, M. and Ricci, M. (1964) *The Films of Marilyn Monroe* New Jersey, Citadel.

Cottrell, R. (1965) 'I Was Marilyn Monroe's Doctor' *Ladies Home Companion* January.

Coward, R. (1984) *Female Desire: Women's Sexuality Today* London, Paladin.

Cowie, E. (1978) 'Woman as Sign'. *m/f* 1.

Curran, J. (ed.) (1982) *Mass Communication and Society* London, Edward Arnold.

Delaney, J., Lupton, M. and Toth, E. (1976) *The Curse: A Cultural History of Menstruation* New York, Dutton.

Dennis, N., Henriques, F. and Slaughter, C. (1964) *Coal is Our Life* London, Tavistock.

Derrida, J. (1981) *Dissemination* Chicago, University of Chicago Press.

de Dienes, A. (1986) *Marilyn Mon Amour* London, Sidgwick & Jackson.

Dougherty, J.E. (1976) *The Secret Happiness of Marilyn Monroe* Chicago, Playboy.

Dunn, J. (1985) 'Identity, modernity and the claim to know better' *Rethinking Modern Political Theory* Cambridge, Cambridge University Press.

Durgnat, R. (1967) *Films and Feelings* London, Faber and Faber.

―――― (1974) 'Mth. Marilyn Monroe' *Film Comment* March–April.

Dyer, R. (1982) *Stars* London, British Film Institute.

―――― (1987) *Heavenly Bodies* London, British Film Institute.

Ellman, M. (1970) *Thinking About Women* New York, Harcourt, Brace Jovanovich.

Evans, P. (1981) *Peter Sellers: the Mask behind the Mask* London, Severn House.

Ewen, S. and E. (1982) *Channels of Desire: Mass Images of the Shaping of the American Consciousness* New York, McGraw Hill.

Ewing, E. (1978) *Dress and Undress: A History of Women's Underwear* London, Batsford.

Fonteyn, M. (1976) *Autobiography* London, Star Books.

Foucault, M. (1982) *The History of Sexuality* Harmondsworth, Penguin, vol. 1.

Franklin, J. and Palmer, L. (1953) *The Marilyn Monroe Story* New York, Rudolph Field.

Freedland, M. (1985) *Jack Lemmon* London, Weidenfeld & Nicolson.

French, B. (1978) *On the Verge of Revolt* New York, Frederick Ungar.

Freud, E.L. (ed.) (1970) Letter of 31 May 1986 *The Letters of Sigmund Freud to Arnold Zweig,* trs E. and W. Robson-Scott, New York, Harcourt Brace World.

Gadamer, H–G. (1986) *The Relevance of the Beautiful* Cambridge, Cambridge University Press.

Gass, W. H. (1977) *On Being Blue: A Philosophical Inquiry* Boston, Godine.

Geertz, C. (1973) *The Interpretation of Cultures: Selected Essays* New York, Basic Books

Genette, G. (1982) *Figures of Literary Discourse* Oxford, Basil Blackwell.

Giddens, A. (1987) *Modern Social Theory and Sociology* Cambridge, Polity.

Gide, A. (1891) Le traite de Narcisse, *Pretexts* New York, Borchardt.

_____ (1982) *Journals 1889-1949* Harmondsworth, Penguin.

Giglio, T. (1956) *Marilyn Monroe* Bologna, Guanda.

Gilson, R. (1969) *Marilyn Monroe 1926-1962* Paris, Anthologie du Cinéma.

Goffman, E. (1971) *The Presentation of Self in Everyday Life* Harmondsworth, Penguin.

Goode, J. (1963) *The Story of the Misfits* Indianapolis, Bobbs-Merrill.

_____ (1986) *The Making of the Misfits* New York, Limelight Editions.

Griffith, R. (1970) *The Movie Stars* New York, Doubleday.

_____ (1971) *The Talkies* New York, Dover Publications.

Guber, S. (1986) ' "The Blank Page" and the Issues of Female Creativity' in E. Showalter (ed.) *The New Feminist Criticism* London, Virago.

Guiles, F. L. (1969) 'Marilyn Monroe' *This Week* 2 March.

_____ (1971) *Norma Jeane: The Life of Marilyn Monroe* London, Panther.

_____ (1985) *Norma Jeane: The Life and Death of Marilyn Monroe* London, Granada.

Hamblett, C. (1966) *Who Killed Marilyn Monroe?* London, Leslie Frewin.

Handel, L. A. (1950) *Hollywood Looks at its Audience* Urbana, University of Illinois Press.

Harris, T. B. (1957) 'The Building of Popular Images: Grace Kelly and Marilyn Monroe' *Studies in Public Communication* 1.

Haskell, M. (1974) *From Reverence to Rape* Harmondsworth, Penguin

Haspiel, J. R. (1975) 'Marilyn Monroe: The Starlet Day' *Films in Review* January–June.

Heath, S. (1982) *The Sexual Fix* London, Macmillan.

_____ (1984) 'Male Feminism. *Dalhousie Review* 64(2).

Hecht, B. (1954) Interview with Monroe *Empire News* (London) 2 May–1 August.

Heidegger, M. (1971) *Poetry, Language, Thought* New York, Harper & Row.

Hembus. J. (1973) *Marilyn Monroe, die Frau des Jahrhunderts* Munich, Wilhelm Heyne Verlag.

Hollander, A. (1978) *Seeing Through Clothes* New York, Viking.

Horkheimer, M. and Adorno, T. W. (1982) 'The Culture Industry:

Enlightenment as Mass Deception' in J. Curran (ed.) *Mass Communication and Society* London, Edward Arnold.

Houston, P. (1963) *The Contemporary Cinema 1945–1963* Baltimore, Penguin.

Hoyt, E. P. (1967) *Marilyn: The Tragic Venus* London, Robert Hale.

Hudson, J. A. (1968) *The Mysterious Death of Marilyn Monroe* New York, Volitant.

Hudson, L. (1982) *Bodies of Knowledge* London, Weidenfeld & Nicolson.

Huston, J. (1981) *An Open Book* New York, Ballantine.

Hutchinson, T. (1982) *The Screen Greats: Marilyn Monroe* New York, Exeter.

Jameson, F. (1981) *The Political Unconscious: Narration as Socially Symbolic Act* London, Methuen.

Jarvie, I. C. (1970) *Towards a Sociology of the Cinema* London, Routledge.

Joyce, A. (1957) 'Marilyn at the Crossroads' *Photoplay* July.

Kael. P. (1973) Review in *The New York Times Book Review* 22 July.

Kahn, R. (1987) *Joe and Marilyn: A Memory of Love* London, Sidgwick & Jackson.

Kanin, G. (175) *Hollywood* London, Hart Davis.

Katz, E. and Lazarsfeld, P. (1955) *Personal Influence* Glencoe, Ill., Free Press.

Kauffman, S. (1973) 'Landmarks of Film History: Some Like It Hot' *Horizon* 15(1), winter.

Kelley, K. (1986) *His Way* New York, Bantam Books.

Kraus, K. (1986) *Half-Truths and One-and-a-Half-Truths* London, Carcanet.

Kristeva, J. (1977) *About Chinese Women* London, Marion Boyars,

───── (1982) *Powers of Horror* New York, Columbia University Press.

Kyrou, A. (1972) *Marilyn Monroe* Paris, Denoel.

Lacan, J. (1977) *Ecrits: A Selection* New York, Norton.

Lasch, G. (1979) *The Culture of Narcissism* New York, Warner Books.

───── (1984) *The Minimal Self* London, Pan.

Lembourn, H. J. (1977) *Diary of a Lover of Marilyn Monroe* New York, Arbor House.

Levin, M. (1976) *Hollywood and the Great Fan Magazines* London, Ian Allen.

Lewis, C. S. (1985) *A Grief Observed* London, Faber & Faber.

Logan, J. (1972) 'A Memory of Marilyn' *Show* September.

Lowenthal, L. (1961) *Literature, Popular Culture and Society* Englewood Cliffs, Prentice-Hall.

Luijters, G. (ed.) (1986) *Marilyn: A Never-Ending Dream* London, Plexus.

Lytess, N. with Wilkie, J. (n.d.) 'My Years with Marilyn', unpublished manuscript, Zolotow Collection, University of Texas, Austin.

McCann, G. (1985) 'Marilyn: The Dream Lingers' *Marxism Today* December.

_____ (1987) 'Biographical Boundaries: Sociology and Marilyn Monroe' *Theory, Culture and Society* 4(4), September.

McClelland, D. (1985) *Hollywood on Hollywood* London, Faber & Faber.

McCreadie, M. (1973) *The American Movie Goddess* New York, Wiely.

McIntyre, A. T. (1961) 'Waiting for Monroe or Notes from Olympus' *Esquire* March.

Mailer, N. (1973) *Marilyn* New York, Grosset & Dunlap.

_____ (1980) *Of Women and their Elegance* London, Hodder & Stoughton.

de Man, P. (1979) *Allegories of Reading* New Haven, University of Minnesota.

Manvell, R. (1975) *Love Goddesses of the Movies* London, Hamlyn.

Marcuse, H. (1964) *One Dimensional Man* London, Routlege & Kegan Paul.

Marilyn Monroe: Rare Recordings 1948–1962 (1979) Sandy Hook Records.

Martin, P. (1956) *Marilyn Monroe* London, Frederick Muller.

Marx, S. (1975) *Mayer and Thalberg New York, Random House.*

Mayer, J. P. (1948) *British Cinemas and their Audiences* London, Dobson.

Mayersberg, P. (1963) 'The Mortal Goddess' *The Listener* 24 January.

Mellen, J. (1973) *Marilyn Monroe* New York, Pyramid Publications.

Merryman, R. (1962) 'Fame May Go By. . . An Interview' *Life* 3/17 August.

_____ (1966) 'Behind the Myth of Norma Jean' *Life* 4 November.

Miller, A. (1958) 'My Wife Marilyn' *Life* 22 December.

_____ (1964) 'With Respect for her Agony – but with Love' *Life International* 24 February.

_____ (1979) *After the Fall* Harmondsworth, Penguin.

Mitchell, J. and Rose, J. (1982) *Feminine Sexuality: Jacques Lacan and the Ecole Freudienne* London, Macmillan.

Monroe, M. (1974) *My Story* New York, Stein and Day.

Moore, R. and Schoor, G. (1977) *Marilyn and Joe DiMaggio* New York, Manor.

Morin, E. (1972) 'La tragédie de Marilyn' *Les Stars,* 3rd edn, Paris, Seuil.

Mulvey, L. (1975) 'Visual Pleasure and Narrative Cinema' *Screen* 16(3).

Murray, E. with Shade, R. (1975) *Marilyn: The Last Months* New York, Pyramid.

Odets, C. (1962) 'To Whom It May Concern: Marilyn Monroe' *Show* October.

Oppenheimer, J. (1981) *Marilyn Lives!* London, Pipeline Books.

Parr, J. (1983) *P.S. Jack Parr* New York, Doubleday.

Patrick, B. (ed.) (1980) *Marilyn* New York, O'Quinn Studios.

Pepitone, L, and Stadiem, W. (1979) *Marilyn Monroe Confidential* New York Simon & Schuster.

Polhemus, T. (ed.) (1978) *Social Aspects of the Human Body* London, Penguin.

Powdermaker, H. (1950) *Hollywood: The Dream Factory* Boston, Little Brown.

Proust, M. (1983) *Remembrance of Things Past* Harmondsworth, Penguin, 3 vols.

Quart, L. and Auster, A. (1985) *American Film and Society Since 1945* London, Macmillan.

Rauh, O. and J. (1983) 'The Time Marilyn Hid Out at Our House' *Ms* 1 August.

Ricoeur, P. (1984) *Time and Narrative* Chicago, University of Chicago Press, vol. 1.

Robinson, D. and Kobal, J. (1974) *Marilyn Monroe: A Life on Film* New York, Hamlyn.

Rollyson, C. E. (1986) *Marilyn Monroe: A Life of the Actress* Ann Arbor, UMI Research Press.

Rorty, R. (1987) 'Liberal Hope and Private Irony,' a Clark Lecture for Trinity College Cambridge, 10 February.

Rosen, M. (1974) *Popcorn Venus* New York, Avon Books.

Rosten, N. (1980) *Marilyn: A Very Personal Story* London, Millington.

Rosten, N. and Shaw, S. (1987) *Marilyn Among Friends* London, Bloomsbury.

Russell, J. (1986) *An Autobiography* London, Sidgwick & Jackson.

Ryan, M. P. (1975) *Womanhood in America* New York, Harper & Row.

Sartre, J-P. (1964) *Les Mots* Paris, Gallimard.

_____ (1976) *Nausea* Harmondsworth, Penguin.

Sayre, N. (1974) 'Screen: Monroe Myths' *New York Times* 1 February.

Schickel, R. (1962) *The Stars* New York, Dial

_____ (1974) *His Picture in the Papers* New York, Charterhouse.

Schjeldahl, P. (1967) 'Marilyn: Still Being Exploited?' *New York Times* 17 December.

Schopenhauer, A. (1978) *Essays and Aphorisms* Harmondsworth, Penguin.

Schwenger, P. (1985) *Phallic Critiques* London, Routledge & Kegan Paul.

Schweppenhauser, H. (ed.) (1971) *Theodor W. Adorno zum Gedachtnis* Frankfurt, Suhrkamp.

Sciacca, T. (1976) *Who Killed Marilyn?* New York, Manor.

Shattuck, R. (1964) *Proust's Binoculars* London, Chatto & Windus.

Shaw, S. (1979) *Marilyn Monroe in the Camera Eye* New York, Hamlyn.

Shelton, R. (1986) *No Direction Home* London, New English Library.

Showalter, E. (1983/4) 'Critical Cross-Dressing: Male Feminists and the Woman of the Year' *Raritan* III(2).

Sichtermann, B. (1986) *Feminity, The Politics of the Personal* Cambridge, Polity.

Signoret, S. (1979) *Nostalgia Isn't What It Used to Be* London, Grafton Books.

Sinyard, N. and Turner, A. (1979) *Journey Down Sunset Boulevard* Ryde, BCW Publishing.

Skinner, Q. (1969) 'Meaning and Understanding in the History of Ideas' *History and Theory* 8.

Skolsky, S. (1954) *The Story of Marilyn Monroe* New York, Dell.

Slatzer, R. (1974) *The Curious Death of Marilyn Monroe* new York, Pinnacle.

Smith, M. (1971) *Marilyn* New York, Barven Publications.

Sontag, S. (1978) *On Photography* London, Allen Lane.

Spada, J. and Zeno, G. (182) *Monroe: Her Life in Pictures* London, Sidgwick & Jackson.

Speriglio, M. (1982) *Marilyn Monroe: Murder Cover-Up* Van Nuys, California, Seville Publishing.

Speriglio, M. with Chain, S. (1986) *The Marilyn Conspiracy* London, Corgi.

Starr, K. (1985) *Inventing the Dream: California Through the Progressive Era,* Oxford, Oxford University Press.

Steinberg, C. (1982) *Reel Facts: The Movie Book of Records* New York, Vintage.

Steinem, G. (with George Barris) (1986) *Marilyn* New York, Henry Holt.

Steiner, G. (1984) 'To Civilize our Gentlemen' *George Steiner: A Reader* Harmondsworth, Penguin.

_____ (1985) *Language and Silence* London, Faber & Faber.

Stern, B. (1982) *The Last Sitting* London, Orbis.

Strasberg, S. (1980) *Bittersweet* New York, Putnum's.

Stuart, A. (1975) 'Reflections of Marilyn Monroe in the last fifties picture show' *Films and Filming* 21(10), July.

Summers, A. (1985) *Goddess: The Secret Lives of Marilyn Monroe* London, Gollancz.

_____ (1986) *Goddess* (revised edition) London, Gollancz.

Tanner, T. (1987) 'Review of Roth' *The Listener* 12 March.

Taylor, C. (1985) *Human Agency and Language* Cambridge, Cambridge University Press.

Taylor, R. G. (ed.) (1983) *Marilyn on Marilyn* London, Omnibus Press.

_____ (1984) *Marilyn in Art* London, Elm Tree Books.

Theweleit, K. (1987) *Male Fantasies,* trs, Stephen Conway, Cambridge, Polity, vol. 1.

Tholen, G. C. (1986) *Wunsch-Denken* Kassel, Gesamthocschule Kassel.

Thomson, D. (1982) 'Baby Go Boom!' *Film Comment* September–October.

Trilling, D. (1964) 'The Death of Marilyn Monroe' *Claremont Essays* New York, Harcourt, Brace World.

Tudor, A. (1974) *Image and Influence* London, Allen & Unwin.

Turin, M. (1976) 'Gentlemen Consume Blondes' *Wide Angle* 1, spring.

Veyne, P. (1971) *Comment en ecrit l'histoire* Paris, Seuil.

Wagenknecht, E. C. (ed.) (1969) *Marilyn, A Composite View* Philadelphia, Chilton Book Co.

Warhol, A. and Hackett, P. (1980) *Popism: the Warhol '60s* New York, Harcourt, Brace Jovanovich.

Warner, M. (1985) *Monuments and Maidens* London, Weidenfeld & Nicolson.

Waters, H. (1972) 'Taking a Look at MM.' *Newsweek* 16 October.

Weatherby, W. J. (1976) *Conversations with Marilyn* London, Robson Books.

Welsch, J. R. (1978) *Film Archetypes: sisters, mistresses, mothers and daughters* New York, Arno Press.

Whitcomb, J. (1960) 'Marilyn Monroe – The Sex Symbol versus the Good Wife' *Cosmopolitan* December.

White, H. (1972) 'The Structure of Historical narrative' *Clio* 1.

_____ (1978) *Tropics of Discourse* Baltimore, John Hopkins.

Widener, D. (1977) *Lemmon: A Biography* London, Allen & Unwin

Wiesel, E. (1981) *Night* Harmondsworth, Penguin.

Williams, D. (1953a) 'Should Marilyn Tone It Down?' *The Mirror* 13 February.

_____ (1953b) 'Marilyn Wants to Turn It Off' *The Mirror* 10 March.

Wills, G. (1982) *The Kennedy Imprisonment* New York, Pocket Books.

Wilson, E. (1985) *Adorned in Dreams* London, Virago.

Wilson, T. (1962) *New York Times* 16 August.

Wiltshire, B. (1982) *Role Playing and Identity: The Limits of Theatre Metaphor* Bloomington, Indiana University Press.

Wittgenstein, L. (1980) *Culture and Value,* trs. Peter Winch, Oxford,

Basil Blackwell.

Wollheim, R. (1984) *The Thread of Life* Cambridge, Cambridge University press.

Woodward, B. and Bernstein, C. (1976) *The Final Days* New York, Avon Books.

Woolf, V. (1955) *To the Lighthouse* New York, Harcourt, Brace Jovanovich.

_____ (1981) *A Room of One's Own* London, Grafton Books.

Yorke, R. (1970) Interview with John Lennon *Rolling Stone* 28 June.

Zolotow, M. (1961) *Marilyn Monroe* London, Panther.

_____ (1973) 'Marilyn Monroe's Psychiatrist' *Chicago Tribune* 16 September.

_____ (1979) 'Joe and Marilyn: The Ultimate L.A. Love Story' *Los Angeles* February.

_____ (1985) 'Memories of Marilyn Monroe' *Reader's Digest* December

Index